The Mathematical Foundation of Multi-Space Learning Theory

This book explores the measurement of learning effectiveness and the optimization of knowledge retention by modeling the learning process and building the mathematical foundation of multi-space learning theory.

Multi-space learning is defined in this book as a micro-process of human learning that can take place in more than one space, with the goal of effective learning and knowledge retention. This book models the learning process as a temporal sequence of concept learning, drawing on established principles and empirical evidence. It also introduces the matroid to strengthen the mathematical foundation of multi-space learning theory and applies the theory to vocabulary and mathematics learning, respectively. The results show that, for vocabulary learning, the method can be used to estimate the effectiveness of a single learning strategy, to detect the mutual interference that might exist between learning strategies, and to predict the optimal combination of strategies. In mathematical learning, it was found that timing is crucial in both first learning and second learning in scheduling optimization to maximize the intersection effective interval.

The title will be of interest to researchers and students in a wide range of areas, including educational technology, learning sciences, mathematical applications, and mathematical psychology.

Tai Wang is a professor affiliated with the Faculty of Artificial Intelligence in Education at the Central China Normal University, China. His research interests include educational technologies, internet psychology, and natural language processing. One major topic of his research is learning environment constructions.

Mengsiying Li is a PhD candidate at the National Engineering Research Center for E-Learning, Central China Normal University, China, whose research focuses on learning spaces, strategies, and behavior.

The Mathematical Foundation of Multi-Space Learning Theory

Tai Wang

Mengsiying Li

Routledge
Taylor & Francis Group

LONDON AND NEW YORK

First published in 2024
by Routledge
4 Park Square, Milton Park, Abingdon, Oxon OX14 4RN

and by Routledge
605 Third Avenue, New York, NY 10158

Routledge is an imprint of the Taylor & Francis Group, an informa business

British Library Cataloguing-in-Publication Data
A catalogue record for this book is available from the British Library

ISBN: 9781032705477 (hbk)
ISBN: 9781032707020 (pbk)
ISBN: 9781032707013 (ebk)

DOI: 10.4324/9781032707013

Typeset in Minion
by codeMantra

To Our Families.

Contents

Figures

Tables

Preface

WE ARE LIVING IN A TECHNOLOGY-RICH WORLD, which not only brings conveniences but also brings confusions, i.e., which technical solution is the best? In different contexts, "best" has different meanings. In learning sciences, it may mean higher retention rate, longer retention duration, or any other effectiveness-related measurements.

In this study, multi-space learning is defined as a micro procedure of human learning which may take place in more than one space, with an implicit or explicit pursuit for effective learning strategies, and such effectiveness can be measured by the indicators based on the knowledge retention duration. The learning procedure is modeled as a temporal sequence of concept learning, incorporated with well-known principles and empirical evidences. And matroid is introduced to reinforce the mathematical foundation of multi-space learning theory.

The aforementioned approach is applied to vocabulary learning (assuming a forgotten word will not affect a word to be remembered) and math learning (assuming a forgotten concept will affect a concept to be mastered), respectively. In vocabulary learning, it was found that matroid-related methods can be used to estimate a single learning strategy's effectiveness, to detect the mutual interference that might exist among learning strategies, and to predict the optimal combination of learning strategies. In math learning, from the matroid-related perspective, it was found that timing is vital both in the first time learning and the second time learning (i.e., reviewing), and this is because scheduling optimization focuses on maximizing the effective interval.

It is expected that this original contribution can discover a new field for learning sciences.

Dr. Tai Wang conceived the main idea, proposed the framework of this book, and wrote each chapter, including developing algorithms and coding. Ms. Mengsiying Li, a PhD student of the former, collected the empirical results presented in Table 5.1 in Chapter 5 and did the first round of proofreading. Ms. Li translated Dr. Tai Wang's original Chinese manuscript of Chapters 1, 2, and 4 into English, though only around 50 lines are kept by Dr. Tai Wang in the final version. She will test multi-space learning theory further again in her thesis, through massive practical mobile vocabulary learning data.

The authors would like to thank the encouragement from the dean of the Faculty of Artificial Intelligence in Education (AIED), Prof. Sanya Liu; the concern from the former dean of the School of Computer Science, Prof. Tingting He; and the advice on how to cultivate a PhD student from a senior professor, Dr. Jixin Wang. The authors would also like to thank Dr. Zhi Liu for sharing the special issue on learning analytics and personalized support across spaces and Dr. Leyuan Liu for his insightful proof of Theorem 5.1 in Chapter 5. They are all at Central China Normal University.

In addition, the authors would like to thank Editor Lian Sun, Editorial Assistant Xiaoyin Feng, Ms Ying Li of the Taylor & Francis Group, and Project Manager Ms. Assunta Petrone of production team in CodeMantra. Their professional work and considerate services are VERY appreciated.

Last but not least, the authors would like to thank Prof. Xiangen Hu's introduction to Prof. Jean-Claude Falmagne, who is one of the two founding fathers of Learning Spaces theory and who has generously granted Dr. Wang to translate their book into Chinese; Prof. Arthur C. Graesser, who summarized learning principles in his inaugural editorial as Editor-In-Chief of *Journal of Educational Psychology*; Prof. Dietrich Albert, who passionately participated in the periodical online/offline forum on Advanced Learning Theories, Applications and Impacts; Prof. Ryan S. Baker, who was one of the initiators of the International Educational Data Mining Society and one of the organizers who held the AIED conference in the authors' hometown, so that the authors can either discuss with them in person or contact them by email. This is definitely multi-space learning for the authors, and the authors learn professional spirits from these great pioneers in more than one space.

The authors believe that one of the best approaches to requite all the above nice people is an original contribution to academic communities. Also, the authors welcome any comments from readers.

The work presented in this monograph is also a staged work of a key project of the National Natural Science Foundation of China: Research on Key Techniques of College Students' Personalized Learning and Intelligent Educational Services under the Multi-space Fusion (No. 61937001).

CHAPTER 1

Introduction on Multi-Space Learning

IN THE LONG-TERM EVOLUTION OF EDUCATION empowered by artificial intelligence (AI), especially in the epidemic, the online-merge-offline (OMO) business model has been widely accepted and commonly adopted in the current education field (Jiang et al., 2022). Nowadays, the mixed scenes of online and offline learning are quite popular in colleges; in a smart classroom (Shi et al., 2003; Zhang, Li, Zhu, Dong, & Hao, 2019; Saini & Goel, 2019; Kwet & Prinsloo, 2020; Kaur, Bhatia, & Stea, 2022), learners use laptops to search scaffolding resources online for the offline instructional material they may not quite follow. At this moment, from a learner-centered perspective, learners are in a state where two different spaces are in a "superposition": their bodies are in an offline space and their ears are still receiving the teacher's voice in the offline space, but their thinking activities in brains focus on the scaffolding resources they are searching in the online space. On the other side, when teachers need to interact with learners, there are also two spaces waiting for them to join in: either offline, i.e., using traditional face-to-face question and answering, or online, i.e., using instant chat tools (even taking part in scrolling comments near the screen of a live lecture) and any other connecting technologies.

The online space not only temporarily satisfies teachers' need of interacting with learners but also becomes part of a newly emerged and popular instructional scene: teachers set up live webcasts in traditional classrooms. This scene may probably be accelerated by the epidemic to be popular and remains active till today. In this scene, the teacher not

DOI: 10.4324/9781032707013-1

only interacts with offline learners but also interacts with online learners located out of the brick-wall classroom. In other words, as a symmetrical counterpart of a learner, a teacher can also be in a two-space superposition. Teachers' being in two spaces can extend the inspiring effect of teacher-learner dialogues in one space into the crowd who observe this dialogue in the other space, encouraging more learners to think and obtain more feedbacks.

Although the above two-space learning scenarios are relatively rare in primary and secondary schools, it is still possible to carry out learning activities in two or more spaces in ordinary schools, whose learning environments are not technology rich. For example, a middle school teacher may organize a discussion in a brick-wall classroom and asks learners to dig deeper online back home. Such learning activity crosses two separate spaces (if a museum visit is added to the homework, there are three separate spaces involved all together), and the whole learning procedure can thus be considered as completed in a multi-space learning environment, from an end-to-end life span point of view.

1.1 DEFINITIONS ON MULTI-SPACE LEARNING

At the beginning of this section, multi-space learning is informally defined as a micro procedure of human learning which may take place in more than one location.[1] Such location may include physical world (offline) and virtual world (online). The physical world includes brick-wall classrooms, museums, concert halls, stadiums, etc. The virtual world includes webinars, live lectures, educational client apps, etc. Since learners are always at some location at some moment, their space-time trajectories can be used to record the learning procedure of an individual, which is also one of essentials of xAPI (Berg, Scheffel, Drachsler, Ternier, & Specht, 2016; Sottilare, Long, & Goldberg, 2017; Manso-Vázquez, Caeiro-Rodríguez, & Llamas-Nistal, 2018; Hernandez, Johnson, & Werkenthin, 2020). In addition, learners can be at more than one location at the same time (besides the aforementioned OMO business model applied in college campus, virtual/augmented reality technologies can make learners feel that they are in different locations simultaneously). However, at a specific moment, the attention of a learner can only stay in one space, even if the

1 At the end of this section, a formal definition will be presented after elaborating through multiple considerations on naming.

learner is in an aforementioned superposition state. The way to figure out in which space this learner is may probably be to ask the learner directly, which is very similar to Schrödinger's Cat. In this book, we are not going to focus on the superposition state itself either in education science or in quantum physics, but to emphasize a fundamental research topic: which space or combinations of spaces are best for learning?

Before we go deep into this topic, please let us explain more about the considerations on the naming of "multi-space learning": by its initial definition, it seems appropriate to name the micro procedure of human learning as "multi-location" learning, then why do we name it "multi-space" learning? The considerations can be summarized that "space" contains more meanings we need than "location."

The first consideration is that both *Merriam-Webster Dictionary* and *Collins Dictionary* provide several precise definitions on space, which are consistent with principles valued by learning science. *Merriam-Webster* explicitly lists "the distance from other people or things that a person needs in order to remain comfortable"[2] as one sub-meaning of "space", and *Collins* explains "personal space"[3] in this way: "if you need your personal space, you need time on your own, with the freedom to do something that you want to do or to think about something." We all agree that either "comfortable" or "freedom" is necessary for effective learning.

More importantly, *Collins* also lists the definitions of "space" in British English and American English.[4] One definition of "space" in British English is: freedom to do what a person wishes to for his or her own personal development. And one definition of "space" in American English is: if you give someone space to think about something or to develop as a person, you allow them the time and freedom to do this. The definitions of "space" in both British English and American English given by *Collins* not only emphasize again the free will that space brings in but also highlight the purpose of entering such space: individual's development. It is well known that "development" has its psychological connotation where learning is an essential approach. And such free will and purpose of development coincide Thomas (1980)'s disposition theory on learning strategies, which brings our second consideration.

---

2 https://www.merriam-webster.com/dictionary/space#dictionary-entry-1.
3 https://www.collinsdictionary.com/dictionary/english/personal-space.
4 https://www.collinsdictionary.com/dictionary/english/space.

> It seems reasonable to assert that the spontaneous use of learning strategies is a matter of disposition: the disposition to perceive a learning task as controllable, to feel responsible for the outcome, and to search actively for ideas for solving the problem posed by the task.
>
> Thomas (1980, pp. 235–236)

The second consideration is as follows. Since it is a fashion to say that education is empowered by AI, as mentioned at the very beginning of this chapter, the specific empowering path seems like: AI $\rightarrow$ learning environment $\rightarrow$ learners. However, the perception that learners are empowered passively by AI in a learning environment is actually caused by the fact that it is the learners who select AI-empowered learning strategies actively and reversely change the surrounding environment. For example, a learner puts down the paper textbook and turns to full-text search to look up a term in the electronic version. It is the learner that alters the physical world around him or her into a virtual world. Therefore, learning strategies selected by the learners are the essentials of a space for learning. In other words, to choose a space for learning is to choose a strategy for learning. In the former example, if learners choose to read a paper textbook, they actually choose a learning strategy: the paper-pencil model of note taking or simply reading. And if learners choose to read an e-textbook, they may want to take the advantage of full-text searching, which is certainly a different learning strategy. Even in a same word-locating learning task, locating them by the indexes at the back of a paper book and locating them by the full-text searching function of an e-book are different learning strategies, not only because they have different cognitive mechanisms but also because they have different effectiveness; since the indexes at the back of a paper book are limited, learners may have to recall some non-target keywords in the indexes as anchors when the words they really want to locate are not in the indexes, rather than directly input any words they want to locate in a full-text searching window.

Therefore, the second consideration actually adds one connotation into the term "space" in our topic: since a learning strategy is always applied within a certain space and shapes a micro space for learning, we actually discuss the effectiveness of strategies when we discuss the effectiveness of multi-space learning. Or, the effectiveness of multi-space learning is represented through the strategies applied in these spaces. And indeed, in the rest of our book, spaces for learning[5] and learning strategies are used interchangeably. Then, why do we not name our topic "multi-strategy learning"? Here comes the third consideration.

The third consideration is that though spaces for learning and learning strategies are used interchangeably in our book, and the latter is used even far more often than the former in Chapters 5 and 6, we do not focus on the internal psychological or pedagogical process of a learning strategy, but on its effectiveness exhibited externally, particularly on its effectiveness on knowledge retention. Therefore, space is more accurate than the strategy on summarizing the scope of our topic.

The fourth consideration is that although the initial definition of multi-space learning originates in learning at multiple locations, the word "space" itself has a meaning: a period of time. Without doubts, learning, like other behaviors, occurs in a certain place and lasts for a certain period of time. In our topic, we concern about how much time a learner needs to take on learning a concept with a learning strategy and how long the learned concept can be retained after applying this learning strategy. In other words, effectiveness is usually about the cost and achievement, and the cost and achievement of effective learning can be measured by time.

The fifth consideration is that the "space" in "multi-space learning" coincides with the "spacing" in the spacing principle. This principle supported by empirical evidences reminds learners to have an appropriate rest between two learning tasks, which is very important for effective learning. Such coincidence implies that "spacing" is part of the models for learning procedure in forthcoming chapters of this book.

To sum up, the *formal* definition on multi-space learning is a micro procedure of human learning which may take place in more than one space, with an implicit or explicit pursuit for effective learning strategies, and such effectiveness can be measured by the indicators based on the duration of knowledge retention.

1.2 DIFFERENCES BETWEEN MULTI-SPACE LEARNING AND OTHER TERMS

Before multi-space learning, there are two similar terms in academic literature, namely, blended learning and learning across spaces. In

[5] To show our respect to an already existed theory: Learning Spaces (Falmagne & Doignon, 2011), we do not use the term "learning space" in our theory, and their work is in fact also quite different from ours.

this section, their differences from multi-space learning are discussed separately.

1.2.1 Multi-Space Learning vs. Blended Learning

Multi-space learning is considered as a procedure of learning which occurs in multiple spaces. Similarly, blended learning is considered as a combination of relatively simple learning components. For example, Alammary, Sheard, and Carbone (2014) summarized three different design approaches for blended learning: low impact (adding extra online activities to a traditional face-to-face course), medium impact (replacing some of the face-to-face activities by online components), and high impact (building the blended course from scratch).

Multi-space learning aims at effective learning, especially knowledge retention. Similarly, Lai, Lam, and Lim (2016) proposed two design principles for blended learning, namely, the consolidation and the extension. "The consolidation principle emphasizes the combination of different components so that students can consolidate their knowledge by engaging in different types of activities. The extension principle emphasizes the extension of learning from one space to another, and that the face-to-face and online components are complementary and have their own strengths and weaknesses." And Lai et al. (2016) proposed that face-to-face and online learning should be "thoughtfully integrated." We believe that their "thoughtful integration" ought to include their concerns about learning outcomes when integrating face-to-face and online learning. And as early as 2000s, Hofmann (2006) recommended that instead of looking at an entire course, the instructor needs to look at each single course learning outcome. And for each outcome, the instructor needs to determine the best delivery option of that outcome.

These above reviews on design approaches and principles for blended learning show that blended learning is teacher-centered, or at least largely since a self-regulated learner can also adopt blended learning as instructors' feedbacks quoted in Lai et al. (2016). While multi-space learning seems learner-centered since it highlights learners' free will on choosing spaces for learning. However, the essential difference between multi-space learning and blended learning does not lie in learner-centered or teacher-centered, but lie in the granularity of "course." Multi-space learning considers learning as a narrowed learning procedure in which concrete knowledge unit (e.g., English words or mathematical concepts) is learned one after another. On the contrary, blended learning focuses on a "normal" course that may involve so many other knowledge indirect-relevant designs (e.g., learning atmosphere), and it is rather difficult to perform a fine assessment on the effectiveness of each learning component weaved. This is our explanation that can be added to the Hofmann's issue list for her assertion that blended learning had not yet been working in general at her time (Hofmann, 2006, p. 30), and we believe this is one of the last few roadblocks for blended learning initiatives.

1.2.2 Multi-Space Learning vs. Learning Across Spaces

The phrase "learning across spaces" discussed here is a term in empirical studies whose abstract level is between describing practitioners' practice and elaborating summarizations or rational reflections on cultural phenomena (e.g., from speculative philosophy). One example of the former is "learning spaces" in practitioners' everyday language. Interested readers may go to two web pages hosted by Western International School of Shanghai, China, either to check their western-style learning facilities[6] or to read their practitioners' thoughts.[7] One example of the latter is (Brown, 2005). Brown (2005) concluded that "no single magic formula will guarantee successful learning spaces on every campus," and "learning space design is a large-scale, long-term project." He suggested that the starting point for rethinking learning spaces begins with an underlying vision for the learning activities these spaces should support, and such vision should be informed by learning theory, institution's specific culture, organizational structure, and fiscal circumstances as well. Although such works are not in the scope of the following discussion, they enrich our comprehension on learning across spaces.

To the best of our knowledge, two journals organized a special issue about empirical studies on learning across spaces. The journal *Digital Education Review* organized a special issue titled "Learning across Settings and Time in the Digital Age" by Esteban-Guitart, Coll, and Penuel (2018). And the journal *User Modeling and User-Adapted Interaction* organized a special issue titled "Learning Analytics and Personalised Support across Spaces" by Martinez-Maldonado, Hernández-Leo, and Pardo (2019).

---

[6] https://www.wiss.cn/wiss-campus-facilities.html.

[7] https://wiss.cn/news/how-can-learning-spaces-improve-learning-.html.

Esteban-Guitart et al. (2018) organized 13 papers, from scholars in Australia, Chile, Finland, Norway, Spain, and USA. Instead of quoting the editorial preface in this special issue, we rephrased a brief summary from the angle of differentiating learning across spaces from multi-space learning.

One major difference between this group of papers and multi-space learning we proposed was that the topics on which these 13 papers discussed related to macro learning procedures. For example (from the most macro learning procedure to the least macro procedure, same as below), in terms of perspectives or philosophies on understanding educational changes, Jornet and Erstad (2018) focused on dialectical relations between continuity and transformation that characterize learning lives. Kumpulainen, Kajamaa, and Rajala (2018) summarized three patterns of agency(capacity of individuals to make their own free choices)and structure (social arrangements that influence or limit the choices) dynamics in students' social activity data collected over one semester.

In terms of intrinsic psychological process, Allen, DiGiacomo, Van Horne , and Penuel (2018) leveraged a longitudinal data set over three years to understand the initial driver behind a young person entering an activity. They found that interest was often not such driver, and instead, other perceived benefits were. García-Pastor (2018) examined 51 digital text of identity produced by learners in English as a second language and suggested more equitable learning spaces because of their reinforced identities as an intercultural speaker.

In terms of digital literacy, Ramirez, Hernández-León, Figueroa-Sandoval, and Aillon-Newman (2018) emphasized the role of art in learning social sciences in higher education since this role is now based on the ability of multimodal platforms that offer access to vast image repositories, which could be utilized to promote better understanding among students about their own world. Cortés, García-Pernía, Fuente, Martínez-Borda, and Lacasa (2018) outlined supporting strategies to generate digital literacy among young people who participated in social networks.

In terms of learning trajectories, this special issue not only included a study on learning trajectories of learners but also included two studies on learning trajectories of teachers. Puigcercós (2018) pointed out that the learners' access to opportunities across or within countries was remarkably unequal, asking for a more inclusive educational system. González-Patiño (2018) and Hernández-Hernández, Sancho-Gil, and Domingo-Coscollola (2018) studied learning trajectories of teachers. The former aimed to promote the use of standard tools of the social web that enabled participants to create knowledge before, during, and after the sessions; as well as disseminate, debate, and recover it afterward, incorporated with the pedagogical approaches of Emmi Pikler. The latter used cartographies as spaces of inquiry to explore of teachers' nomadic learning trajectories.

In terms of who involved in spaces for learning, Willis and Exley (2018) and Levinson and Barron (2018) both believed that parents or families could be involved. The former provided an example of the potential of digital technologies to mediate the usual disconnect between home and school settings to encourage parent-school engagement, which invited future, longer-term research based on the findings. The latter did not concentrate on parent-school engagement, but on Latino immigrant families' use of technology to learn, as well as on how digital resources served to help children and parents explore content across settings. One practical purpose of such innovations was to expand parents' roles as collaborative learners of English as a second language.

In terms of in-school and out-of-school, Engel, Coll, Membrive, and Oller (2018) revealed that the rate of participation in information and communication technology mediated activities outside school was low, and these activities were associated with limited learning experiences. Castells and Illera (2018) provided an example of trans-media named *The Ancestral Letter*. It offered us a more understandable and effective way of coordinating between the school-based and informal settings and between the digital setting and everyday life.

Different from a grand vision on macro learning procedures reflected by the 13 papers published in the special issue organized by Esteban-Guitart et al. (2018), Martinez-Maldonado et al. (2019) organized four papers that prefer micro learning procedures. Their authors were from Canada, Ireland, Spain, and the United States. In their preface, Martinez-Maldonado et al. (2019) defined learning across spaces as "learning is happening in an increasing number of spaces, at different times and with the support of different tools and technologies," and "from a pedagogical perspective, this combination of tools, resources, and settings presents the potential of new more effective learning experiences." Such definition and note are similar to ours; however, each study published in this special issue had its distinct understanding on the framework of learning across spaces.

For example, when Azcona, Hsiao, and Smeaton (2019) predicted the performances on learning programming languages based on log files, they assumed learning management system and programming tool were two different digital spaces. When Zhu, Xing, Costa, Scardamalia, and Pei (2019) used multiple sensors to study knowledge construction on the dimensions of emotion and cognition, they assumed physical spaces and digital spaces were two different spaces. When Hutt et al. (2019) embedded eye trackers from a laboratory to the laptops in a classroom to detect the mind-wandering moments, they assumed learning had crossed from a laboratory space to a classroom space. When Amarasinghe, Hernández-Leo, and Jonsson (2019) integrated log files from multiple pedagogical tools to predict learning engagement, they assumed personal learning space and collaborative learning groups were two kinds of learning spaces. These examples illustrate diverse understandings on learning across spaces; nevertheless, we can also observe that the framework of learning across spaces followed by these examples is not yet unified. It is partly because the concept of learning across spaces is developing as the studies are ongoing, so that the granularity of spaces for learning is not consistent.

In the following chapter, a different division on spaces for learning is proposed, based on which the mathematical foundation of multi-space learning theory is constructed.

CHAPTER 2

Partition Spaces to Optimize Learning Effectiveness

THERE are many ways to divide complex spaces for learning into simple or single ones, level by level. At the first level, a complex space for learning can be partitioned by examining two states of three roles in it, namely, teachers (a human being or an artificial one), learners (more than one or just one), and network connections between teachers and learners (existing or not). Then, the $2 \times 2 \times 2 = 8$ spaces

TABLE 2.1 An Example of Partitions on Spaces for Learning at the First Level

Label	Nature of the Teacher	Net. Conn. b/w Teach. and Learn.	Number of Learners	Typical Scenario
Space 1	Human	No	Multiple	Traditional classroom
Space 2	Human	No	One	Offline individual tutoring
Space 3	Human	Yes	Multiple	MOOC[a]/ SPOC[b]
Space 4	Human	Yes	One	Online individual tutoring
Space 5	Artificial	Yes	Multiple	Computer-supported online

[a]Massive open online courses.
[b]Private closed online courses.

DOI: 10.4324/9781032707013-2

for learning at the first level and their typical scenarios are listed in Table 2.1.

2.1 PARTITION SPACES INTO DEEPER SUB-SPACES

From Spaces 1 to 7, they exhibit an approximate trend on the time dimension: from the farthest to the latest. One of the latest progresses in Space 6 is the ChatGPT [1] born in November 2022. Although it introduces a lot of controversy, we are looking forward to using it as a reliable reference after it is able to guarantee its answers' authorities. Space 7 has been already realized in some science museums, though sometimes it acts like a high-end version of a "walkman" (a mobile device with multiple location-sensitive, automatically recorded playbacks). A more cutting-edge version of Space 7 must be a humanoid tutor equipped with lightweight and authority-guaranteed ChatGPT. Space 8 in Table 2.1 can be seen as an end of a long evolution from Space 1 when a human teacher is replaced finally by a humanoid tutor. The recent progresses in Space 6 and Space 7 make Space 8 look so real that it may come true in the next morning. However, we believe technologies cannot lose the control of human beings. Therefore, human ethics or beliefs on education will work at some time, so that the tension between technology first and human first will find its own balance eventually in the educational field.

After partitioning spaces at the first level (named "teacher-learner interaction level") as Table 2.1, we can continue to partition each space from Spaces 1 to 7 into more concrete levels by following the revised Bloom's taxonomy (L. W. Anderson & Krathwohl, 2001). The second level would be educational objectives, such as cognitive, affective, and psychomotor.

In the second level, the cognitive educational objective can be further divided into six layers, namely, remembering, understanding, applying, analyzing, evaluating, and creating.

In the third level, the knowledge for remembering can be further divided into four categories, namely, factual, conceptual, procedural, and meta-cognitive.

Therefore, if the granularity of other counterparts (e.g., attitudes for responding) can be as concrete as the above four categories of knowledge for remembering (e.g., respect or disrespect), the number of combinations of the revised Bloom's educational objectives and single

[1] https://openai.com/blog/chatgpt

space for learning can be at the same scale of 3 domains × 6 layers × 4 categories of knowledge for remembering×7 single spaces for learning = 504 sub-spaces. Or, we may have sub-spaces for learning at the scale of 500 (i.e., each combination can be considered as a sub-space for learning). This is a top-to-bottom approach. And we believe that such approach will not end in dividing spaces into more concrete ones at the scale of 500, when the learning content and the learning strategies are not determined.

Following a bottom-to-top approach, we argue that each concrete educational objective determines one or more learning strategies. As the formal definition for multi-space learning in Chapter 1, a concrete concept or knowledge unit and one of its learning strategies adopted by the learner are two major factors in a micro learning procedure (what is going to be learned and how to learn it) in this book. More specifically, we aim at the remembering level of factual knowledge (category) in the cognitive process (domain). And we believe that similar reasoning presented in later chapters can also be generalized to other combinations or at least shed light on paradigms to handle them.

2.2 PROBLEM DEFINITION: OPTIMIZING LEARNING EFFECTIVENESS

Because we consider spaces for learning as essential in learning strategies, one of the considerations behind our formal definition on multi-space learning in Section 1.1, and because we are affirmative on the prediction of Martinez-Maldonado et al. (2019) that "from a pedagogical perspective, this combination of tools, resources, and settings presents the potential of new more effective learning experiences," *optimizing learning effectiveness* is the primary problem of our multi-space learning theory proposed to solve.

Mathematically, an ordered pair of integers $< \ concept, space \ >$ (interchangeably $< concept, strategy >$) denotes that a learner learns concept c in space s (interchangeably, a learner adopts strategy s to learn concept c). Suppose that a learner L spends time T on learning knowledge K. K consists of concept $c_1, c_2, , \ldots, c_i, , \ldots, c_n$ $(i = 1, 2, \ldots, n)$. Suppose that it will take L time $t(i, j)$ on learning if L decides to learn c_i in s_j. $\mathbf{I}(t, j)$ indicates that whether or not L chooses to learn c_i in s_j, i.e.,

$$\mathbf{I}(i, j) = \begin{cases} 1 & \text{learning } c_i \text{ in } s_j \\ 0 & \text{else} \end{cases} \tag{2.1}$$

And,

$$T = \sum_{i=1}^{n} \sum_{j=1}^{m} t(i,j)\mathbf{I}(i,j) \tag{2.2}$$

Learning effectiveness can be measured by different parameters. Learning duration, i.e., to find $\arg\min_{i,j} T$, is one of them. And maximum retention duration (but not defined here) is also another one. To solve this kind of optimization problems, we need to introduce the Matroid theory.

CHAPTER 3

Matroid Theory

3.1 BRIEF INTRODUCTION

Matroid comes from an abstract on the linear independence in a vector space. In linear algebra, once a set of vectors is linearly independent, it obtains two properties, namely, hereditary property and exchange property.

The hereditary property is that the subset of a set of linearly independent vectors is also linearly independent, i.e., the linear independence is "inherited" from the set to the subset.

The exchange property is a little bit complicated. It refers to a mathematical phenomenon: if there are two sets of linearly independent vectors (S_1 and S_2), and S_1 has less vectors than S_2, then there exists at least one vector e in the difference set $S_2 - S_1$, which can guarantee that the new set $S_1 \cup \{e\}$ is also linearly independent.

This phenomenon is trivial in linear algebra, or it is actually a well-known theorem in linear algebra. What we concern about is not how we re-prove this theorem, but look at what happens in a different perspective. It can be found that in this phenomenon, S_2 can "send" an e to S_1, so that the new vector set $S_1 \cup \{e\}$ is also linearly independent (i.e., the property of linear independence is passed like an apple to $S_1 \cup \{e\}$ from S_2). The vector set $S_1 \cup \{e\}$ can "return" e to $S_2 - \{e\}$, so that the "new" vector set S_2 is linearly independent (i.e., the property of linear independence is again passed like an apple to S_2 from $S_1 \cup \{e\}$). In other words, the property of linear independence is "exchanged" between $S_1 \cup \{e\}$ and S_2.

Whitney (1935) retained the hereditary and exchange properties, but removed the specific content (e.g., the linear independence in the above

DOI: 10.4324/9781032707013-3

example) that was inherited and exchanged, and named any ordered pair of a limited set and its subset collection which exhibit both hereditary property and exchange property as *matroid*. Such naming is reasonable, since a matrix is probably the first professional terminology coming into one's brain that meets these two properties. Note that each row or column in a matrix is equivalent to a vector of a set in the above example.

Many definitions are proposed and proved equivalent since Whitney's initial definition on matroid. One of the commonly used, rigorous, and formalized definitions is as follows (J. G. Oxley, 1992, pp. 7–8).

Definition 3.1. *A matroid M is an ordered pair $(E, \mathcal{I})$ consisting of a finite set E and a collection $\mathcal{I}$ of subsets of E satisfying the following three conditions:*

(M1) $\emptyset \in \mathcal{I}$.

(M2) *If $I \in \mathcal{I}$ and $I' \subseteq I$, then $I' \in \mathcal{I}$.*

(M3) *If I_1 and I_2 are in $\mathcal{I}$ and $|I_1| < |I_2|$, then there is an element e of $I_2 - I_1$, such that $I_1 \cup \{e\} \in \mathcal{I}$.*

The condition (M2) is the aforementioned hereditary property (West, 2001, p. 349), and (M3) is the aforementioned exchange property (Welsh, 2010, p. 14, 7) minus 0.3em or independence augmentation axiom (J. G. Oxley, 1992, p. 8).

3.2 APPLICABILITY FOR MULTI-SPACE LEARNING

Section 3.1 briefs that matroid abstracts two properties from linear independence and does not require the description on what is inherited or exchanged in its generalized definition. However, if we cannot find anything that can be both inherited and exchanged in multi-space learning, it is certainly impossible to apply matroid theory in multi-space learning. Fortunately, the first thing that comes to mind that fits matroid's definition is knowledge. It not only meets the hereditary property and exchange property literally (no one will deny that knowledge can be both inherited and exchanged) but also meets matroid's formal definition. Below are the details.

3.2.1 Course Knowledge Can Be Modeled by Matroids

Suppose E is a finite set of all concepts in a meaningful and close course, such as Geometry in a secondary school. E is usually printed in the form of indexes at the back of a textbook. $\mathcal{I}$ appears in the form of a table-of-contents page that covers the scope of the course semantically. Not all the concepts in E are printed on the table-of-contents page; however, they are all included separately by the scope of chapters whose titles are listed in the table of contents ($\mathcal{I}$ is a collection of subsets of E). $\emptyset$ corresponds to any margin place in the table-of-contents (Condition **(M1)** is satisfied or we can make $\emptyset$ as a special concept in E).

Therefore, if a chapter I is in the textbook (denoted by $\mathcal{I}$), its sections or subsections (I') are also included in $\mathcal{I}$ (Condition **(M2)** is satisfied).

And, if a chapter I_1 contains less concepts than another chapter I_2, due to the dense connections among Geometric concepts, we can always find one concept or a group of concepts[1] (i.e., e) in $I_2 - I_1$, so that the combination of I_1 and e is still in the textbook (Condition **(M3)** is satisfied). Therefore, the Geometry course meets matroid's definition. In addition, if we replace concepts with words and replace chapters with units, modeling a language course by matroid will also pass the similar examination.

The above examinations on modeling course knowledge by matroid reveal that matroid is good at describing a partitioning procedure when an object can be divided into multiple components and sub-components while retaining its nature unchanged[2], and endless subdivision theoretically (there will be an end in practice). Course knowledge, in fact, is not the only thing that meets matroid's definition in multi-space learning theory. Space for learning is the second one.

A space for learning can be divided into more concrete ones like course knowledge being subdivided endless (theoretically) above. And one of such subdivisions is exemplified in Section 2.1. It is easy to examine that this example also meets matroid's definition. However, as emphasized in Section 1.1 (i.e., the second consideration on multi-space learning's definition), simply sitting in a space for learning without passively or actively adopting any strategies to learn cannot make learning happen. So, our key point is not only to confirm that learning spaces are indeed matroids but also to examine whether learning strategies also meet matroid's definition.

---

[1] This is one of essential differences between our theory and Knowledge Spaces Theory or Learning Spaces Theory (Doignon & Falmagne, 2016).

[2] Not all the subdividable objects are matroids. It depends on how to subdivide. A pile of flour can be divided into balls of flour, and subdivided into pinches of flour... They are all flour. In this way, flour is a matroid. However, if a ball of flour is divided chemically, i.e., sugar, fat, protein, vitamin,..., and later into chemical elements such as Calcium, Iron, etc. flour is not a matroid.

At the first glance, it seems that there is no reason that learning strategies are not matroids; after all, sub-strategies for learning (if subdividable) are also strategies for learning. However, a rigorous examination cannot be absent.

3.2.2 Learning Strategies Can Be Modeled by Matroids

This subsection validates that learning strategies are matroids by empirical evidences. Empirical evidences supporting *(1) learning strategies are subdividable; (2) learning strategy effectiveness can be measured in multiple ways; and (3) learning strategy can be identified by empirical approaches* are organized in a near chronological order.

3.2.2.1 Learning Strategies Are Subdividable

Mayer (1988, p. 11) defined learning strategies as "behaviours of a learner that are intended to influence how the learner processes information." As early as 1988, McCombs had divided skill domains into meta-cognitive, cognitive, and affective strategies (p. 141). To provide a research agenda for learning and motivational strategies training, she not only proposed to combine these three learning strategies together but also advanced a generative model of underlying processes and skills, including attentional strategies, retrieval strategies, rehearsal strategies, and self-management strategies (p. 155).

Since then, more and more studies work on sort out learning strategies. For example, Oxford (1990, p. 16) distinguished two classes of language learning strategies (LLS), namely, direct and indirect. In direct LLS, there are three groups, namely, memory, cognitive, and compensation. Indirect LLS also have three groups, namely, meta-cognitive, affective, and social. Ultimately and altogether, she subdivided learning strategies into 2 classes, 6 groups, 19 sets, and 62 subsets (intentionally so written!) and organized them in a tree form (p. 17). As an instance, "taking notes" is one of 62 leaves in this tree. It is one of three subsets of "creating structure for input and output" set, which is one of four sets of cognitive strategies group (p. 21).

Entering this century, rather than proposed by experienced scholars from a top-to-bottom approach, the subdividable structure of learning strategies naturally emerged from a questionnaire (Dowson & McInerney, 2004) on learning strategies among 720 middle ($n = 602$) and senior ($n = 118$) school students, which was a semi bottom-to-top approach. Confirmatory factor analysis on the responses to this questionnaire validated a hierarchical structure among learning strategies. The root levels are cognitive and meta-cognitive. The cognitive strategies are rehearsal (including listing, memorizing, reciting, and/or naming facts/items to be learned), elaboration (including paraphrasing, generating analogies, and reviewing previous work), and organization (including selecting, sequencing, outlining, reordering, or summarizing important information). The meta-cognitive strategies are planning (including prioritizing, time management, scheduling, setting realistic goals, and arranging work environments appropriately), monitoring (including self-checking for understanding, self-testing, and organizing reviews of learned material), and regulating (including attempting different ways to learn material, seeking explanations from teachers, and identifying mistakes in reasoning). Therefore, similar to Oxford (1990, p. 16)'s taxonomy, Dowson and McInerney (2004)'s hierarchical structure also has four levels.

In Jeng, Wu, Huang, Tan, and Yang (2010)'s review on add-on impact of mobile applications in learning strategies, more concrete strategies integrated into ubiquitous learning apps came into readers' sight, e.g., knowledge awareness map recommending the best matched materials according to learners' tasks and current physical locations (El-Bishouty, Ogata, & Yano, 2007), presenting the appropriate English news to learners based on the fuzzy item response theory (Chen & Hsu, 2008).

3.2.2.2 Learning Strategy Effectiveness Can Be Measured in Multiple Ways

In the recent decade, learning strategy studies not only continue aiming at more concrete strategies but also pay more attention on the effectiveness of learning strategies by empirical experiments.

On the effectiveness, Glogger, Schwonke, Holzäpfel, Nückles, and Renkl (2012) utilized learning journals (236 participants in 9th grade of medium-track high schools) to assess learning strategies. They found that quality and quantity of cognitive strategies predicted learning outcomes, controlling for prior knowledge. Learners who combined cognitive and meta-cognitive strategies were particularly successful. Learners who mainly used one type of strategy performed similarly poorly as did learners who hardly used strategies. Broadbent and Poon (2015) investigated the effectiveness of 10 self-regulated learning strategies for higher education (including combined strategies, meta-cognition, time management, effort regulation, peer learning, elaboration, rehearsal, organization, critical thinking, and help seeking) by systematically reviewing 12 empirical studies filtered from 1789 papers. They found that the strategies of time management, metacognition, effort regulation, and critical thinking were positively correlated with academic outcomes, whereas rehearsal, elaboration, and organization had the least empirical support. Peer learning had a moderate positive effect; however, its confidence intervals crossed zero. In addition, the effects of strategies in traditional face-to-face settings became weaker when they were generalized to online context.

On the concrete strategies, McConnell et al. (2017) examined the effectiveness of 11 common active learning strategies most of which were more concrete than those in Glogger et al. (2012). Their effectiveness was ranked by a total strategy score yielded by a combination of strategy utility and learning efficacy scores. They were peer instruction, think-pair-share, minute papers, lecture tutorials, concept maps, concept sketches, case studies/problems, jigsaw, teaching with models, gallery walks, and role playing, from the highest strategy score to the lowest strategy score.

Davis, Chen, Hauff, and Houben (2018) turned to adopt more diverse indicators to measure the effectiveness of learning strategies, not just learning outcomes as in Glogger et al. (2012) nor research validation rubric (strength of evidence from low to high) and strategy utility rubric (divided into pre-class preparation, student strategy use, and task characteristics) as in McConnell et al. (2017). Davis et al. (2018)'s 14 indicators of learning strategy effectiveness included final grade, completion rate, learning gain, exam score, long-term retention, learning transfer, ontrackness, engagement, forum participation, video engagement, revision, persistence/coverage, self-efficacy, and efficiency. These indicators were more empirical, concrete, goal-oriented, and learner-centered. They adopted the number of empirical experiments reporting either positive or null or negative effects on these indicators, to rank the learning strategies in the context of learning at scale (i.e., MOOC). They identified three most promising strategies for effectively leveraging learning at scale, namely, cooperative learning, simulations & gaming, and interactive multimedia (other seven learning strategies were mastery learning, metacognitive learning, questioning, spaced vs. massed practice, matching learning styles, feedback, and programmed instruction). Although these learning strategies were relatively abstract than those in Glogger et al. (2012) and McConnell et al. (2017) and some might overlap with another ones, the voting-from-experiments approach adopted for comparison seemed fairer. Nevertheless, such fairness might come from the inclusive definitions on these ten online learning strategies, which prevented readers from knowing more about the effectiveness of sub-categorical learning strategies, though this was not their primary goal (their primary goal was to find out promising common characteristics of online learning strategies).

In the last five years, Sinnayah, Rathner, Loton, Klein, and Hartley (2019) found that a combination of three very concrete active learning strategies (guided-inquiry learning in small group-based tutorials; use of audience response systems in large group lectures; and use of pre-tutorial quizzes before weekly tutorials) improved student academic outcomes in first-year paramedic bioscience. However, they could not clearly state which of the active learning activities were responsible for the clearly evident increase in student grades and a reduction in fail rates.

Besides formal learning in classroom settings, Kortsch, Schulte, and Kauffeld (2019) found four informal learning strategies for craft workers (informal learning from oneself, from others, from other sources, and from new media). Craft workers used three patterns of strategy combinations, namely, balanced high, person-oriented, and balanced low. These three combination patterns differed in intensity and combination with the learning strategy use. More intense combination patterns were positively related to internal marketability[3] but were not related to external marketability.

3.2.2.3 Learning Strategy Can Be Identified by Empirical Approaches

Besides identifying learning strategies by learning journals in Glogger et al. (2012), Catrysse et al. (2018) used eye-tracking to distinguish very strategic students, characterized using cognitive processing and regulation strategies, from low strategic students, characterized by a lack of cognitive and regulation strategies.

Since (1) learning strategies are subdividable, (2) learning strategy effectiveness can be measured in multiple ways, and (3) learning strategy can be identified by empirical approaches, learning strategy meets matroid's definition. Its validation procedure can imitate course knowledge's validation procedure. Since learning strategy is a matroid, many properties a

---

[3] How valuable one is to the current employer (internal marketability) or to other employers (external marketability).

matroid owns can be used to search the optimal combination of learning strategies in a specific knowledge learning setting.

3.3 MATROID AND ANTIMATROID

Matroid and antimatroid are not a pair of antonyms as their names look like. In this section, a clarification is made from our perspective with the help of multiple professional references. Dietrich (1989) listed a commonly used definition on antimatroid as follows.

Definition 3.2. *An antimatroid A is an ordered pair $(E, \mathcal{I})$ consisting of a finite set E and a collection $\mathcal{I}$ of subsets of E satisfying the following three conditions:*

(A1) $\emptyset \in \mathcal{I}$.

(A2) *If $I \in \mathcal{I}$ and $I \neq \emptyset, \exists$ an element $e \in I$ s.t. $I - e \in \mathcal{I}$.*

(A3) *If I_1 and I_2 are in $\mathcal{I}$ and $I_1 \nsubseteq I_2$, then there is an element e of $I_1 - I_2$, such that $I_2 \cup \{e\} \in \mathcal{I}$.*

Some symbols are replaced to make the semantic differences between Definition 3.1 and Definition 3.2 more highlighted while the original connotation remains unchanged. The semantic differences between these two definitions are compared one by one.

First, Condition **(M1)** in matroid's definition is the same with Condition **(A1)** in antimatroid's definition.

Second, Condition **(M2)** in matroid's definition says that if a set belongs to the collection, its subset does as well. While Condition **(A2)** in antimatroid's definition says that if a set belongs to the collection, there exists one of its subsets that miss only one element also belongs to the collection. Therefore, Condition **(A2)** is more stringent than Condition **(M2)**. And Condition **(A2)** $\Longrightarrow$ Condition **(M2)**.

Third, Condition **(M3)** in matroid's definition says that if there are two sets and one is larger than the other, then an element can be selected from the part of the larger set that exceeds the smaller one, so that this selected element and the smaller set can consist of a new set which still belongs to the collection. While Condition **(A3)** in antimatroid's definition says that it is not necessary to identify which set is larger or smaller, only if one set is not included by the other, an element can thus be selected from the part of the latter set that exceeds the former one, so that this selected element and the latter set can consist of a new set which still belongs to the collection. Therefore, Condition **(A3)** is weaker than Condition **(M3)**. And Condition **(M3)** $\Longrightarrow$ Condition **(A3)**.

Although it is rather difficult to trace who first names "antimatroid" and why uses a name which is inconsistent with general intuition, it is believed that such "anti-" prefix refers to the "inconsistent" entailment relationships between the two pairs of corresponding conditions.

Since either pair of conditions (i.e., **(M2)** vs. **(A2)**, and **(M3)** vs. **(A3)**) is not mutually opposed, there is a chance that an intersection of matroid and anti-matroid exists. In fact, Korte and Lovász (1988) proposed a general operation to produce a greedoid from a matroid and an antimatroid on the same ground set, which we will not brief here to reduce the burden of reading. Dietrich (1989) gave another example which is a matroid and also an antimatroid. Dietrich (1989) and Korte, Lovász, and Schrader (1991, p. 124) also pointed that matroids are actually the interval greedoids without lower bounds, and antimatroids are the interval greedoids without upper bounds. Korte et al. (1991, p. 126) further validated that the intersection of a matroid and an antimatroid need not always yield a greedoid. If the intersection is a greedoid, then it is also the meeting of a matroid and an antimatroid. A detailed application of antimatroid can be found in Section 4.1.1.1.

3.4 BENEFITS OF BEING MATROIDS

Being a matroid can bring multiple benefits. Before we exemplify these benefits, we illustrate our visions on the course graph construction.

Figure 3.1 shows a 3D "building" for geometric concepts in elementary textbooks designed by the Ministry of Education (M.O.E.) in China. Each "floor" is a semester. "S6th" indicates spring semester in the 6th grade, while "A1st" indicates autumn semester in the 1st grade. The connected concepts indicate that they are in a same unit. The left concepts usually appear ahead of the right ones. The majority of structures is total-sub, while a few of them are progressive.

Figure 3.1 exhibits a horizontal relationship between concepts, particularly in a same semester ("floor"). Figure 3.2 exhibits a vertical relationship between concepts. To visualize more clearly, concepts are organized into clusters rather than displaying individually: locations & directions (1), units of length or weights (2), angles (3), observing objects (4), area and its units (5), graphics (6), and motion of graphics (7), from easiness to difficulty.

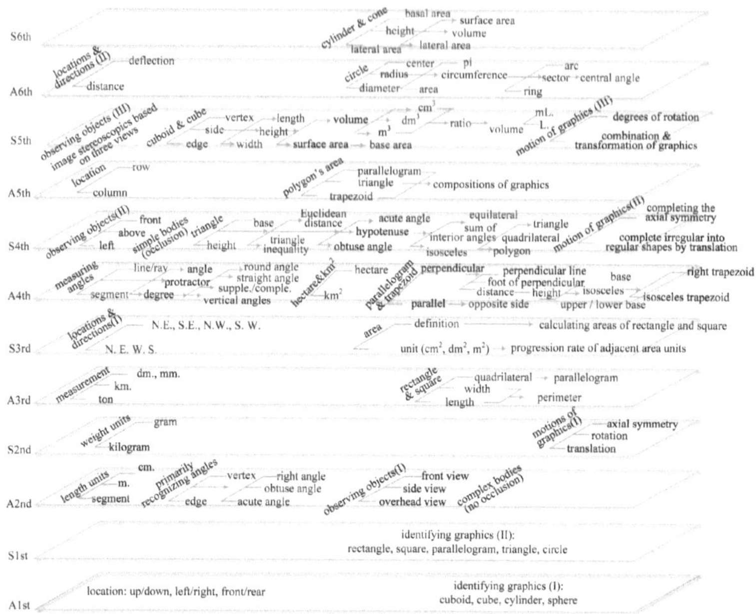

FIGURE 3.1 A 3D building for geometric concepts in elementary textbooks (M.O.E. version).

In the other dimension of the $x - y$ plane, the number of concepts in a cluster is recorded. The semester dimension (vertical) stays the same as Figure 3.1. The format of coordinates is cluster index, number of concepts, and semester.

So, each point in Figure 3.2 indicates the index of concept cluster and the number of concepts in this cluster in a certain semester. A complete list is presented in Table 3.1.

In Figure 3.2, the number of connections between a pair of clusters is the number of overlapped concepts across the two clusters. Overlapped concepts are defined as the shared connotations between concepts that surpass 40% of the larger one in Chinese.

For example, (1,3,1) denotes that in autumn semester of the first grade, the cluster of "locations and directions" contains three concepts ("上下", up/down; "前后", front/rear; "左右", left/right). (4,4,3) denotes that in autumn semester of the second grade, the cluster of "Observing Objects" contains four concepts ("前面", front view; "上面", overhead view; "左

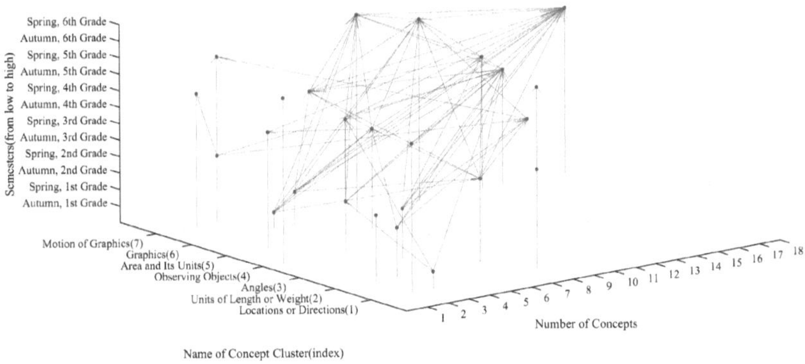

FIGURE 3.2 A 3D map of concept cluster connections in elementary textbooks of geometry (M.O.E. version).

面", side view; "复杂形体", complex bodies). And (1,3,1) has an arrow to (4,4,3) whose weight is 3 (omitted in the figure) because "上下"(up/down) and "上面"(front view), "前后"(front/rear) and "前面"(front view), and "左右"(left/right) and "左面"(side view) are three pairs of overlapped concepts. The threshold parameter, 40%, is set by empirical experience in Chinese language, e.g., ruling out this pair: "长方体" (cuboid) and "正方形" (square). Other languages can also develop their own threshold parameters, and careful readers may already have found that the "side view" and "left" do not have anything shared in their names, which is probably due to the different naming conventions among languages. This issue can be solved even manually when an individual believes a pair of concepts have close relationships with each other.

3.4.1 Example 1 of Benefits: Accelerating the Construction of a Key Concept Network

A course graph as in Figure 3.2 may have a small portion of concept nodes which have significantly more connections than other ones. In pedagogy, these key concepts are usually either basic and commonly used concepts whose out-degrees are above the average and usually parts of foundations for a course, or highly integrated ones whose in-degrees are above the average and usually require deep understanding or comprehensive skills. If teachers can notice the special importance of these nodes during course

TABLE 3.1 A Complete List for Geometric Concepts in Elementary Textbooks

Semester[a]	Cluster Index	Concepts
A1st	1	Up/down, front/rear, right/left
A1st	6	Cuboid, cube, cylinder, sphere
S1st	6	Rectangle, square, parallelogram, triangle, circle
A2nd	2	cm., m., segment
A2nd	3	Vertex, edge, right angle, obtuse angle, acute angle
A2nd	4	Front view, overhead view, side view, complex bodies
S2nd	2	Gram, kilogram
S2nd	7	Axial symmetry, rotation, translation
A3rd	2	dm, mm., km., ton
A3rd	6	Rectangle, square, quadrilateral, length, weight parallelogram, perimeter
S3rd	1	E., S., W., N., N.E., S.E., N.W., S.W
S3rd	5	Coverage, cm^2, dm^2, m^2, area of a rectangle area of a square, progression rate of adjacent area units
A4th	3	Segment, line, ray, angle, protractor, degree, straight angle round angle, supplementary angle, vertical angles, complementary angle
A4th	5	Hectare, km^2
A4th	6	Parallelogram, trapezoid, parallel, perpendicular, perpendicular line foot of a perpendicular, distance, opposite side, height, base, upper base lower base, isosceles, isosceles trapezoid, right angled trapezoid
S4th	4	Front, above, left, abstract/simple bodies
S4th	6	Triangle, base, height, distance, triangle inequality, right triangle hypotenuse, obtuse triangle, acute triangle, isosceles triangle, equilateral triangle, sum of interior angles in triangle, sum of interior angles of quadrilateral, sum of interior angles in polygon
S4th	7	Completing the axial symmetry complete irregular into regular shapes by translation
A5th	1	Row, column
A5th	5	Parallelogram's area, triangle's area, trapezoid's area compositional graphics' area
S5th	4	Imaging stereoscopies based on three views
S5th	6	Cuboid, cube, vertex, edge, side, length, weight, height surface area, volume, cm^3, dm^3, m^3, base area progression rate of adjacent volume units, capacity, L., m.L.
S5th	7	Degrees of rotation, combination, transformation
A6th	1	East by north, west by south, east by south, west by south north by east, north by west, south by east, south by west
A6th	6	Circle, center of a circle, radius, diameter, circumference, pi area of a circle, ring, arc, sector, central angle
S6th	6	Cylinder, cone, bottom surface, side, height surface area, lateral area, volume

[a] A1st indicates autumn semester in the first grade; S6th indicates spring semester in the sixth grade.

preparation and thus pay more attention to these nodes, or learners can understand the special importance of these nodes during their reviews and thus pay more attention to them, their instruction or self-regulated learning may have a better effectiveness.

An experienced teacher may find the concept network of special importance by professional knowledge accumulated for years; however, a naive teacher does not have the same accurate identification on these nodes as an experienced teacher, nor does a learner. Fortunately, algorithms can help non-experienced persons to identify these elements of special importance, which can be used as a base to form (if personalized adjustment is needed) a concept network of special importance in pedagogical perspective.

In graph theory, a minimum vertex cover can work as the small portion of concept nodes that are of special importance mentioned in the two previous paragraphs. A vertex cover for graph $G(V, E)$ is a subset T of G such that for each edge $(u, v) \in E$, at least one of u or v belongs to T (Pullan, 2009). The minimum vertex cover is a vertex cover that has the minimum number of elements. The minimum vertex cover problem is equivalent to maximum independent set problem (Pullan, 2009; Lozin, 2017). Both of them are NP-hard for a general graph. Nevertheless, it is well known that the Greedy algorithm delivers an optimum solution when $G(V, E)$ is a matroid in a polynomial time (Turner, Ehrgott, & Hamacher, 2015; Chekuri, 2018). When $G(V, E)$ is a matroid, it can have cycles. When $G(V, E)$ has cycles and is also a matroid, it is named a cycle matroid in J. Oxley (2001, p. 11).

A cycle matroid $G(V, E)$ should have cycles when it is presented in the form of a graph, while Figure 3.2 cannot guarantee that any course graph always has cycles, though it is very obvious that learning can have multiple paths. In case there is not a cycle in a course graph, we can make some minor adjustments to the course graph, so that this requirement can be satisfied. For example, note that Figure 3.2 is originally designed for the use of the first time learning, and, therefore, it is directed. However, when it comes to a second time learning, it can be turned into an undirected graph. If this can still not make a cycle appear, we can then add additional connections as few as possible to make a cycle appear because individuals may probably not fully agree on the course graph drawn by automatic running applications without any manual revision. In addition, if there are isolated nodes, we can focus on the rest nodes that are connected and put these isolated nodes aside.

Bazgana, Toubaline, and Tuza (2011) has studied the k most vital nodes independent set (k most vital nodes vertex cover) problem, which consists of determining a set of k vertices whose removal results in the greatest decrease in the maximum weight of independent sets (minimum weight of vertex covers, respectively), given an undirected graph with weights on its vertices. As mentioned previously, the minimum vertex cover problem is equivalent to maximum independent set problem. Contrary to this, however, it follows from their results that for $k \leq 1$ the optimal solutions of k most vital nodes independent set and k most vital nodes vertex cover can be substantially different. Their work lays a sound foundation for those teachers or learners who would like to present how vital a concept is by deletion.

3.4.2 Example 2 of Benefits: Planning Reviews by Minimum Spanning Tree

Figure 3.2 shows a 3D connection map of concept clusters. Although a concept-level map can also be constructed and restored in a matrix, such map is omitted here due to its complexity. However, we can imagine that what if the weight of a connection is replaced from how overlapped two connected concepts is, to how soon it will take the learning/reviewing journey along the connection (path), i.e., the time spent on learning/reviewing the two connected concepts. Naturally, this alteration will lead to an efficiency problem: how soon will it take to review all the concepts? Is there an optimum (i.e., the shortest) path connecting all the concepts?

In matroid theory, the answer to the above questions is minimum spanning tree. The Kruskal's algorithm (Kruskal, 1956) and Prim's algorithm (Prim, 1957) for minimum spanning tree were illustrated in Cormen, Leiserson, Rivest, and Stein (2022, pp. 591–597) in detail, and their relationships with matroid theory[4] was detailed in Cormen et al. (2009, pp. 437, 439–440, 442).

3.4.3 Example 3 of Benefits: To Our Study

Since learning strategies can be modeled as a matroid, according to greedy algorithm, to obtain the minimal T in Equation (2.2), we only have to pick up the minimal time $t_{\min}(i)$ spent on each c_i. That is,

[4] The reason underlying why greedy algorithms are effective at finding minimum spanning trees is that the set of forests of a graph forms a graphic matroid (Cormen, Leiserson, Rivest, & Stein, 2009, p. 642).

$$t_{\min}(i) = \min_{1 \leq j \leq m} (t(i, j)) \tag{3.1}$$

And,

$$T_{\min} = \sum_{i=1}^{n} t_{\min}(i) \tag{3.2}$$

In Chapters 5 and 6, other applications of matroid theory in multi-space learning in a language course domain and a math course domain under different optimal targets will be illustrated, respectively.

Although matroid theory is the foundation stone of multi-space learning theory, other mathematical foundations of existing theories and parameter values from empirical studies are also needed. Such literature not only helps us open our mind but also adjusts our thinking trajectories, and sometimes may even reshape[5] the initial optimal function presented in Equation (2.2) in Chapter 2, i.e., changing the optimal target from minimum learning duration to maximum retention duration. Therefore, the summarization on the current foundations for learning sciences is needed in the forthcoming Chapter 4.

---

[5] Roughly speaking, Chapter 5 considers forgetting and having a rest between two adjacent learning tasks, and Chapter 6 considers dependent relationships between concepts, in addition.

CHAPTER 4

Current Foundations of Learning Sciences

4.1 MATHEMATICAL FOUNDATIONS

4.1.1 Learning-Science-Specific Theories or Technologies

4.1.1.1 Knowledge/Learning Spaces Theory

The knowledge spaces theory or learning spaces theory has three cornerstones, namely, knowledge structure, knowledge spaces, and learning spaces. Their mathematical definitions are cited from Doignon and Falmagne (2016) as follows. Cosyn and Uzun (2009) proved that the learning space defined below is an antimatroid, whose definition can be found in Section 3.3.

Definition 4.1. *A knowledge structure is a pair* $(Q, \mathcal{K})$ *consisting of a nonempty set* Q *and a collection* $\mathcal{K}$ *of subsets of* Q*; we assume* $\emptyset \in \mathcal{K}$ *and* $Q \in \mathcal{K}$*. The set* Q *is called the domain of the knowledge structure* $(Q, \mathcal{K})$*. The elements of* Q *are the items, and the elements of* $\mathcal{K}$ *are the knowledge states, or just the states.*

Definition 4.2. *A knowledge space* $\mathcal{K}$ *is a knowledge structure and is closed under union, i.e.,* $\cup\mathcal{C} \in \mathcal{K}$ *for any subcollection* $\mathcal{C}$ *of* $\mathcal{K}$*.*

Definition 4.3. *A learning space is a well-graded knowledge space. A knowledge space is well-graded if for any two states* K *and* L *in* $\mathcal{K}$*, there exists a natural number* h *such that* $|K \triangle L| = h$ *and a finite sequence of states* $K = K_0, K_1, \ldots, K_h = L$ *such that* $|K_{i-1} \triangle K_i| = 1$ *for* $1 \leq i \leq h$*.*

DOI: 10.4324/9781032707013-4

Translated into a natural language, knowledge structure is a set of knowledge states and their combinations in a course, a knowledge space requires that any combinations of these knowledge states will not be beyond the course scope, and a learning space requires that there is only a state difference between two adjacent knowledge states. (A knowledge state refers to whether or not a learner masters a concept in the original theory. However, if this is difficult for some readers to understand, they can treat knowledge state as concept initially.) If the concepts in a course can form a learning space, according to Knowledge/Learning Spaces Theory, the knowledge state of its learner can be precisely identified among a large amount of possible knowledge states through only a few sophisticatedly designed ask-and-answer interactions with the learner.

Knowledge/Learning Spaces Theory has won a great success in practice, taking ALEKS as an example (Cosyn, Uzun, Doble, & Matayoshi, 2021). However, not all the courses can be modeled as learning spaces, e.g., language courses and art courses. And Knowledge/Learning Spaces Theory does not aim to solve how fast a learner can learn a concept or how long a learner can retain the knowledge newly learned.

4.1.1.2 Item Response Theory

Knowledge/Learning Spaces Theory aims at measuring the knowledge state of a learner. While Item Response Theory aims at measuring the ability level of a learner. Item Response Theory is based on the following assumptions (Hambleton & Swaminathan, 1985).

- Unidimensionality of ability assumption—Only one ability or trait is necessary to "explain" or "account" for examinee test performance (Hambleton & Swaminathan, 1985, p. 16).
- Local independence assumption—An examinee's responses to different items in a test are statistically independent (Hambleton & Swaminathan, 1985, p. 23).
- Item characteristic curves assumption—An item characteristic curve (ICC) is a mathematical function that relates the probability of success on an item to the ability measured by the item set or test that contains it (Hambleton & Swaminathan, 1985, p. 25). The probability of an individual examinee providing a correct answer to an item depends only on the form of the ICC (Hambleton & Swaminathan, 1985, p. 26).

- Speededness-free assumption—The tests are not administered under speeded conditions (Hambleton & Swaminathan, 1985, p. 30).

There are many models describing ICC. Among these models, one commonly used model is three-parameter logistic model. Its mathematical form (Hambleton & Swaminathan, 1985, p. 37) is

$$P_i(\theta) = c_i + (1 - c_i)\frac{e^{Da_i(\theta - b_i)}}{1 + e^{Da_i(\theta - b_i)}} \quad (4.1)$$

where $P_i(\theta)$ is the probability that an examinee with ability level θ answers item i correctly, b_i is the item difficulty parameter, a_i is the item discrimination parameter, D is 1.7 (a scaling factor), and c_i represents the probability of examinees with low ability correctly answering an item and is the lower asymptote of ICC (ibid., p. 38).

In practice, the speededness-free assumption cannot usually hold. The response data not only entail the information to measure the ability that a test originally focuses on but also entail the information of at least one ability, time management, since a practical test is usually constrained by limited time duration. Under a similar consideration, the unidimensionality of ability assumption is often broken through. Even at the time of Hambleton and Swaminathan (1985), they enhanced this assumption by adding "what is required for this assumption to be met adequately by a set of test data is a 'dominant' component or factor that influences test performance. This dominant component or factor is referred to as the ability measured by the test"(ibid., p.17).

Item response theory has also won a great success, especially in the field of ability tests, e.g., Test of English as a Foreign Language (Lalor, Wu, & Yu, 2016) and Graduate Record Examinations (Kingston & Dorans, 1982). Nowadays, multidimensional item response theory (MIRT) has been edited as a handbook (Bonifay, 2020). And the discrimination with unidimensional and multidimensional item response theory models for educational data is discussed in Mahmood and Miller (2022). MIRT is also integrated with deep neural networks to support analytic automated essay scoring (Takumi & Masaki, 2022).

4.1.1.3 Epistemic Network Analysis

Both learning spaces theory and item response theory are used to assess learners' status, either knowledge state or ability level. Knowledge state and ability level are outcomes of learning procedures. Learning procedures such as reasoning can be speculated from the knowledge states or skill levels diagnosed by these two theories, or visually restored by a technology named epistemic network analysis (ENA).

ENA is a computational technology that can "nail" the epistemic elements on a two-dimensional plane, so that the connections between these nailed elements can be visualized and compared on a same base. These epistemic elements used by ENA are codes that label a text either of a dialogue or of a monologue according to a certain framework. For example, "I'm going to report on three things. Listen and fill me in." (Liu, Wang, Bo, & Zhang, 2022) is labeled "Communicating/Scrutinizing," one of eight codes in a scientific reasoning and argumentation framework proposed by Fischer, Kollar, Ufer, Sodian, and J. (2014). The connections among these epistemic elements form a thinking trajectory, presenting that a learner's mind is wandering between the codes. And a group of these trajectories can exhibit a thinking pattern. Readers can read a tutorial on the usage of ENA written by Shaffer, Collier, and Ruis (2016).

ENA was initially proposed by Shaffer et al. (2009). At that time, coded elements were placed on a network graph whose layout was determined by an algorithm (Kamada & Kawai, 1989) under aesthetic standards: not only more closely associated codes appear closer together but also less edge crossings and more uniform distribution of vertices and edges. In this initial version of ENA, however, only the relative positions of nodes are important, not their absolute horizontal or vertical position in the graph. This type of ENA presented a snapshot of how minds travel, but failed to place them on a common base, resulting in different backgrounds of pictures when they essentially share common context.

In 2017, Shaffer and Ruis fixed the position of a code node by an optimization routine to minimize the distance between the centroid of the network graph to the point that represents the network under singular value decomposition (SVD). A more detailed mathematical foundation behind this computational procedure was revealed in Bowman et al. (2021). ENA has now been adopted in more than 300 papers.

4.1.2 General Approaches in Educational Data Mining and Learning Analytics

Besides "simple" general approaches such as clustering, classification, identification, or regression/analysis of variances, a few examples of

TABLE 4.1 Complicated General Approaches Adopted by EDM and LA

Rank of Freq.	EDM	LA
1	Bayesian/probabilistic reasoning	Bayesian/probabilistic reasoning
2	Decision trees	Decision trees
3	Support vector machines /sequential minimal optimization	Logistic regression /survival analysis
4	Artificial neural networks	Hybrid systems/bagging /boosting/fusion
5	Random forests	Support vector machines /sequential minimal optimization
6	Association rules /sequential pattern mining	Random forests

"complicated" [1] general approaches include decision trees, random forests, artificial neural networks, fuzzy logic, support vector machines, and genetic/evolutionary algorithms.

Charitopoulos, Rangoussi, and Koulouriotis (2020) reviewed complicated general approaches adopted by educational data mining (EDM) and learning analytics (LA) communities in the name of soft computing. These complicated general approaches share a common nature: avoiding the exhaustive search of the solutions' space and go for possibly suboptimal solutions, with their heavy reliance on rich but often imprecise, incomplete, or even contradictory data which happen to be one of the characteristics of learning sciences shared with other social sciences. As essentially iterative optimization algorithms, these complicated general approaches are quite a supplement to those theories and technologies above, i.e., learning spaces theory and item response theory aim at optimal solutions to assessments on a learner' s knowledge state and capacity, respectively, while ENA places the nodes by an optimization routine (Table 4.1).

Although top priorities of EDM and LA are not identical in the soft computing aspect, it may be argued that (i) Bayesian/probabilistic reasoning, (ii) decision trees, (iii) support vector machines, and (iv) random forests enjoy an increased popularity in both EDM and LA research.

[1] In fact, when integrated into a toolbox and called by one command line, these "complicated" approaches will also become "simple." We do not intend to discriminate either one of these two kinds of general approaches. Sometimes, "simple" may only imply that they are used quite often or commonly seen.

4.1.2.1 Bayesian/Probabilistic Reasoning

Bayesian/probabilistic reasoning is listed at the top of increasingly popular soft computing approaches in EDM and LA research argued by Charitopoulos et al. (2020). Although they did not mention the reason (or they did because of their observation), we can speculate in our perspective that the reasons may include the following: Bayesian/probabilistic reasoning is probably one of the computing approaches that are almost like human learning, i.e., having a prior knowledge background and updating knowledge with new information. Here is a numeric example of how Bayesian reasoning works, partially translated from a Chinese columnist.[2] A totally different English story named "Asia" can also be found at a tutorial[3] hosted by a private company on Bayesian network software development, norsys.com.

4.1.2.1.1 A Numeric Example Suppose there is a crime scene in a villa. The villa's owner A was beaten to death. B, C, and D are three suspects. There are three facts as follows:

- B was the elder son of A. If A is dead, B will be A's successor. It was B who found A's dead body and locked down the crime scene.
- C was the younger son of A. Since C could not share A's inheritance, C and A were in a tense relationship.
- D was the maidservant of A. Since D looked after A's living, D and A were in a good relationship.

Before a detective investigates the case, based on the above three facts, the detective may have a prior estimation on the probabilities of each person being a killer. Say (k is short for killer):

$$P(k = B) = 1/2, \quad P(k = C) = 1/3, \quad P(k = D) = 1/6.$$

These three probabilities are called prior probabilities. When the detective examined A's body, the first clue came out: the estimated time of death of A was 6 p.m. to 8 p.m. At that time,

B was having a video conference and never left;

[2] https://zhuanlan.zhihu.com/p/100746813, Last access on August, 10th, 2023.

[3] https://www.norsys.com/tutorials/netica/secA/tut_A1.htm#diagnosis, Last access on August, 10th, 2023.

C was playing online video game with a friend, but this friend recalled that C was offline for a while;

and D was cleaning the house, but without an alibi.

Therefore, based the first clue (denoted as Clue 1), the likelihoods of committing a crime at the interval [6 p.m., 8 p.m.] if a suspect is the killer are, for example:

$$P(\text{Clue } 1|k = B) = 0, \ \ P(\text{Clue } 1|k = C) = 1/6,$$
$$P(\text{Clue } 1|k = D) = 1/3.$$

But the detective is more concerned about the posterior probability of each suspect, i.e., when Clue 1 holds, the probability of B, C, or D being the killer. Therefore, we have to follow the definition of conditional probability to calculate, which is,

$$P(k|\text{Clue } 1) = \frac{P(\text{Clue } 1|k)P(k)}{P(\text{Clue } 1)}$$

Since

$$\sum_{k \in B,C,D} P(k|\text{Clue } 1) = 1$$

We have

$$\frac{P(\text{Clue } 1|B)P(B)}{P(\text{Clue } 1)} + \frac{P(\text{Clue } 1|C)P(C)}{P(\text{Clue } 1)} + \frac{P(\text{Clue } 1|D)P(D)}{P(\text{Clue } 1)} = 1$$

So,

$$P(\text{Clue } 1) = \frac{1}{2} \cdot 0 + \frac{1}{3} \cdot \frac{1}{6} + \frac{1}{6} \cdot \frac{1}{3} = \frac{1}{9}$$

Therefore, back to the definition of conditional probability, we have $P(k = B|\text{Clue } 1) = 0$ and $P(k = C|\text{Clue } 1) = P(k = D|\text{Clue } 1) = 1/2$. It means that B's suspect is excluded. Then, the suspects are narrowed down to C or D.

The original version of this example provides an additional clue, Clue 2, to continue the reasoning or inference. But we believe Clue 1 is just enough to illustrate the nature of Bayesian/probabilitic reasoning: re-allocating the probability distributions over events with the help of newly coming clues (or facts, information, knowledge, etc.).

In many cases like this case, some events may have higher probabilities than other events after this kind of iterative re-allocations, so that Bayesian inference can be applied to narrow a large scope of events down to a small range. This nature makes Bayesian/probabilitic reasoning be especially good at inferring the cause behind a phenomenon. The numeric example here is about Bayesian/probabilistic reasoning in crime scene investigation, and the "Asia" story aforementioned is about Bayesian/probabilistic reasoning in disease diagnosis. Without doubt, Bayesian/probabilistic reasoning can also be applied in educational assessment (Culbertson, 2016; Almond, Mislevy, Steinberg, Yan, & Williamson, 2015) and knowledge tracing(Pelánek, 2017; Sarsa, Leinonen, & Hellas, 2022).

4.1.2.2 Decision Trees

Bayesian/probabilistic reasoning excels at tracing backward to infer the reasons, while decision tree is good at looking forward to predict outcomes or classify objects, e.g., student academic performance (Matzavela & Alepis, 2021). Although deep learning can now predict outcomes or classify objects more accurately than traditional algorithms, decision tree still has its unique advantages: it can provide people with explainable procedures of prediction or classification, a.k.a, white box.

A decision tree is a rooted, directed tree akin to a flowchart. Each internal node corresponds to a partitioning decision, and each leaf node is mapped to a class label prediction. Each internal node is programmed with a splitting rule, which partitions the domain of one (or more) of the data's attributes. Based on the splitting rule, the data item is sent forward to one of the node's children, beginning at the root. This testing and forwarding are repeated until the data item reaches a leaf node (Aggarwal, 2014, p. 88).

The splitting criterion plays a critical role in the performance of decision trees. A commonly used decision tree named ID3 adopts entropy as the splitting criterion. Specifically, entropy measures the degrees of chaos. Either classification or prediction is to reduce the degrees of chaos. By the formula of entropy, reducing chaos will decrease the entropy. Therefore, if a node is split, ID3 will select a node whose entropy is the lowest (i.e., the least number of misclassifications) to be the splitted node in the next iteration. However, this splitting criterion has a side effect: the decision tree may grow too deep to pursue the most accurate classification/prediction outcomes, which results in over-fitting, i.e., less generalization when facing new data.

To inhibit this side effect, another commonly used decision tree named C4.5 was proposed to adopt gain ratio as a splitting criterion instead. The gain ratio is a parameter that balances the entropy minimization and the tree's size.

Another influential and popular decision tree program, CART, uses the Gini index for a splitting criterion. This can be interpreted as the expected error if each individual item were randomly classified according to the probability distribution of class membership within each subset (Aggarwal, 2014, p. 94). CART has the same side effect as ID3. However, CART prunes branches that are too long.

More details of ID3, C4.5, and CART, such as their sources and related formulas, can be found in Aggarwal (2014, pp. 93–103).

A quick tutorial named "Golf Today?" on how to use decision trees is illustrated by RapidMiner.[4] Shalin (2018) cited this tutorial as an opening in his article and led this tutorial to analyze online learning data.

In healthcare field, Patricia et al. (2022) compared 12 classifiers on three datasets of human activity recognition. Among four decision-tree based classifiers, the performance order in the light of precision and recall quality metrics from high to low is J48, JRip, LMT(logistic model tree), and RepTree. In the field of learning sciences, to the best of our knowledge, there is one such comparison (Alshehri, Alhakami, Baz, & Alsubait, 2020) close to the scales in Patricia et al. (2022)'s contribution, though Alshehri et al. (2020) did not compare EDM tools and technologies on a same dataset. According to Alshehri et al. (2020)'s contribution, it can be concluded that the performance order in the light of highest accuracy appeared in the references from high to low is J48, JRip, and RepTree, similar to Patricia et al. (2022)'s finding.

4.1.2.3 Support Vector Machines

Noble (2006) tried to explain support vector machines (SVMs) without reading any equations by emphasizing four basic concepts, namely, the separating hyperplane, the maximum-margin hyperplane, the soft margin, and the kernel function. Although he did not formalize an official definition on SVM, we can concatenate these four concepts into a sentence as our perspective on SVM: SVM is an optimization method that tries to separate two or more classes of data points (also called vectors in a high-dimensional

[4] https://docs.rapidminer.com/latest/studio/operators/modeling/predictive/trees/parallel_decision_tree.html, Last access August, 12th, 2023.

space) with a maximum-margin hyperplane; if a hard margin cannot be reached, we can either turn to a soft margin or alter the kernel function instead. A complete and detailed introduction on SVM can be found in Suthaharan (2016).

SVM is effective in a high-dimensional space (e.g., Cardona and Cudney (2019) used this method to predict degree completion within three years by college students), even when the number of features is greater than the number of samples (this is possible in educational data since a learner may have multiple features, e.g., gender, grade, ...,). It is memory efficient and versatile. However, if the number of features is much greater than the number of samples, the method is likely to result in poor performance (Roy, Kar, & Das, 2015).

4.1.2.4 Random Forests

Random forests are a combination of tree predictors such that each tree depends on the values of a random vector sampled independently and with the same distribution for all trees in the forest (Breiman, 2001). Using a more natural language,[5] Random forests grow many classification trees; to classify a new object in the form of an input vector, put the input vector down each of the trees in the forest; each tree gives a classification, and we say the tree "votes" for that class; the forest chooses the classification having the most votes (over all the trees in the forest).

The above descriptions may make random forests look like a merely ensemble of decision trees; however, the beliefs behind these two algorithms are quite different. The idea behind random forests is that a group of people each of whom only has limited knowledge on a specific problem may deliver a better solution as a whole than a senior expert does.[6]

In fact, this idea works. Caruana and Niculescu-Mizil (2006) presented a large-scale empirical comparison among 10 supervised learning methods, on 11 different datasets over 8 different performance metrics. With excellent performance on all eight metrics, they found that random forest not only ranks second but also closes to the first place, while decision tree ranks eighth, and SVM ranks fifth.

---

[5] https://www.stat.berkeley.edu/ breiman/RandomForests/cc_home.htm, Last access on August, 14th, 2023.

[6] The idea behind random forest makes this algorithm be usually categorized into the family of ensemble machine learning, see Cutler, Cutler, and Stevens (2012) for example, rather than an extension of decision tree.

Although random forest has excellent performance on prediction in learning sciences, e.g., (Tian, Huang, Zhang, & Wang, 2021; Leena et al., 2021), currently, its usage in learning sciences is much less frequent than decision tree, according to the statistics by Issah, Appiah, Appiahene, and Inusah (2023). This is probably because random forest scarifies interpretability for prediction accuracy, to some extent.

Recently, researchers at Duolingo Portnoff, Gustafson, Rollinson, and Bicknell (2021) revealed their two new developed systems, namely, Checkpoint Quiz and Review Exercises, which can facilitate learners to choose to sequence their studying and how much they choose to revisit old material. They showed general approaches exemplified above are central to making use of language learning assessment data at scale.

4.1.3 Foundations in Adjacent Disciplines

4.1.3.1 Cognitive Science

Andler (2012) tidied and organized mathematics involved in Cognitive Science. According to his mere sample but not an exhaustive list, there are two levels in his framework, namely, macro-level and micro-level frameworks.

The macro-level framework integrates abstract or pure information-processing theories (also known as computational theories). The mathematics involved in this level include logic(s), signal detection theory, control theory, machine learning, probability theory, game theory, and category theory.

The micro-level framework refers to neural dynamics, which includes linear algebra, statistical physics, tools from advanced analysis (ordinary nonlinear differential equations, partial differential equations, Fourier analysis, and wavelets...) and from dynamical systems theory, and methods in the theory of motor control (differential geometry and kinematics).

Although this whole section focuses on mathematical foundations, it would be less fair to exclude some works only because they are less related to mathematics, e.g., J. R. Anderson (1996)'s adaptive character of thought-rational (ACT-R) theory. In this theory, complex cognition arises from an interaction of procedural knowledge (production rules) and declarative knowledge (chunks). The individual cognitive units are created by simple encodings of chunks or production rules. Both Irfan and Gudivada (2016) and Geary (2021) consider ACT-R as the theory behind MATHia (Ritter & Fancsali, 2016).

4.1.3.2 Mathematical Psychology

Luce (1964) was probably the first scholar in modern time who summarized mathematics used in Mathematical Psychology. Different from the two-level division partially determined by the application areas in Andler (2012)'s mere samples, Luce (1964) divided mathematics used in Mathematical Psychology into five groups, which are partially determined by the nature of mathematics involved.

The first group is fundamentals, including set theory, functions, relations, orderings, axiomatics, and Boolean algebra. The second group is analysis, including calculus, transform theory, and complex variable theory. The third group is probability theory, including elementary probability, stochastic process, and Markov chains. The fourth group is algebra, including elementary algebra, matrices, and concatenation algebras. The fifth group is topology and geometry. At his time, "neither is much used" (Luce, 1964).

In the updated and expanded version of Luce (1964), not only the fifth group is upgraded to Chapter 4 in Batchelder, Colonius, Dzhafarov, and Myung (2016) in the name of "network analysis" due to its frequent usage but also the updated version is already expanded into a three-volume set: (I) foundations and methodology (Batchelder et al., 2016), (II) modeling and measurement (Batchelder, Colonius, & Dzhafarov, 2021), and (III) perceptual and cognitive (Ashby, Colonius, & Dzhafarov, 2023).

The first volume in this three-volume set is more closer to mathematical foundations for learning sciences than other two volumes. Ten chapters in this volume are *selected concepts from probability*; *probability, random variables, and selectivity*; *functional equations*; *network analysis*; *knowledge spaces and learning spaces*; *evolutionary game theory*; *choice, preference, and utility: probabilistic an deterministic representations*; *discrete state model of cognition*; *Bayesian hierarchical models of cognition*; and *model evaluation and selection*. The terminologies in the ten italic chapter names appear more or less in the previous paragraphs.

Therefore, due to the large overlap with senior scholars' summarization on adjacent disciplines (i.e., touching cornerstones of adjacent disciplines' mathematical foundations), we believe we have covered cornerstones in the mathematical foundations for learning sciences, though the subtitles of this chapter are far from a complete list.

4.2 EMPIRICAL STUDIES

4.2.1 Transactional Distance Theory

Transactional[7] distance theory proposed that three factors can affect the learning effectiveness, namely, the content structure of instructional material, the psychological distance between the learner and the instructional material (including the teacher), and the autonomy of the learner (Moore, 1973, 1991). In multi-space learning, when the content structure of instructional material remains unchanged, the psychological distance between the learner and the instructional material will change with the type of space for learning. However, the theory itself does not specify how to measure the content structure or the psychological distance. Transactional distance theory is often used as an explanatory theory.

4.2.2 Cognitive Load Theory

From our understanding, cognitive load theory inherited from transactional distance theory. It is assumed that there are two factors that can affect learning effectiveness, namely, intrinsic cognitive load and extraneous cognitive load (Sweller, 2011).[8] By the following definitions, a genetic relationship can be found between intrinsic cognitive load and content structure of instructional material, and between extraneous cognitive load and the psychological distance between the learner and the instructional material, as well.

In cognitive load theory, intrinsic cognitive load is the complexity of mastering the task set without learning methods for reference, and it depends on the connection between the elements that constitute

---

[7] Transactional here does not refer to transactions in the financial field, but it refers to mutual influence, that is, in a mutually constituted situation, each element restricts and is restricted by another element (Rosenblatt, 1988, p. 2). In instructional activities, teachers and learners have such a mutual influence relationship, e.g., learning can help (i.e., restrict) teaching in turn.

[8] Sweller, the founder of cognitive load theory, once proposed two subcategories of cognitive load in 1988, namely, intrinsic cognitive load and extraneous cognitive load, and then added a third subcategory, namely, germane cognitive load, in 1998. However, Sweller later accepted Kalyuga (2011)'s doubts and suggestions on the necessity of germane cognitive load. Since 2011, he has not mentioned germane cognitive load in his published articles and appropriately adjusted the definitions of intrinsic cognitive load and extraneous cognitive load in Sweller (2011). Although some researchers and some textbooks still use Sweller's previous three-load classification, our summary here adopts his latest two-load classification. Recently, Kirschner, Sweller, Kirschner, and Zambrano (2018) further extended the cognitive load theory (including its two-load classifications) for individual learning into collaborative learning.

the task. Extraneous cognitive load is not only affected by these connections but also affected by the way the instructional materials are presented or the instructional strategies. When the intrinsic cognitive load caused by the instructional materials remains unchanged, the extraneous cognitive load will change with instructional design or learning strategies.

Although cognitive load theory and transactional distance theory look similar in a communication perspective (i.e., to successfully sending a message, how the message is structured and how the message is interpreted are two important factors), one of the forward steps moved by cognitive load theory is that intrinsic cognitive load and extraneous cognitive load are more concrete than their counterparts in transactional distance theory, so that it is relatively easier to measure them theoretically. Therefore, in multi-space learning, whether or not there is an optimal solution to reduce the cognitive load maximally is a worthy direction of future studies. In line with this direction, we assume that the cognitive load or the effectiveness of a learning strategy can be examined by the time consumption on learning or by the knowledge retention duration, practically.

4.2.3 Evidence-Based Principles for Teaching or Learning

In Graesser (2009)'s inaugural editorial for the *Journal of Educational Psychology*, he summarized 7 evidence-based principles for teaching, and 25 evidence-based principles for learning. We listed these 7 teaching principles and their counterparts in learning principles (11 items altogether) in Table 4.2 and listed the other 14 learning principles in related parts in a learning procedure in Table 4.3. Other recent summarizations on evidence-based pedagogical principles include (Overton & Johnson, 2016) and (Mayer, 2021), which are not in a typical stye of academic survey, but in a form of notes for practice. Overton and Johnson (2016) collected several milestones in learning strategy studies both classical and recently built, e.g., taking notes by pen may outperform taking notes by laptop in terms of conceptual understanding (Mueller & Oppenheimer, 2014). Mayer (2021) integrated 14 evidence-based principles to illustrate how to design an effective instructional video.

Both Tables 4.2 and 4.3 show that there are lots of overlaps between teaching principles and learning principles. Even in Table 4.3 where no counterpart in teaching principles can be found seemingly, learning

TABLE 4.2 Seven Evidence-Based Teaching Principles and Their Counterparts in Learning Principles

Teaching Principles[a]	Learning Principles[b]
T1 space learning over time	L01 contiguity effects L05 spacing effect L16 segmentation principle
T2 interleave worked example solutions with problem solving exercises	L10 stories and example cases
T3 combine graphics with verbal descriptions	L03 dual code and multimedia effects
T4 connect and integrate abstract and concrete representations of concepts	L11 multiple examples
T5 use quizzing to promote learning	L04 testing effect L06 exam expectations
T6 help learners allocate study time effectively	L24 self-regulated learning
T7 ask deep explanatory questions	L17 Explanation effects L18 deep questions

[a]Tx denotes the x-th teaching principle in Graesser (2009).
[b]Lx denotes the x-th learning principle in Graesser (2009).

TABLE 4.3 Other 14 Learning Principles and Related Parts in a Learning Procedure

Related Part	Learning Principles[a]
Driving force of learning	L19: cognitive disequilibrium L20: cognitive flexibility L23: discovery learning
Presented learning materials	L02: perceptual-motor grounding L08: organization effects L09: coherence effect L15: manageable cognitive load L25: anchored learning
Homework	L07: generation effect L14: desirable difficulties L21: goldilocks principle
Feedback	L12: feedback effects L13: negative suggestion effects L22: imperfect precognition

[a]Lx denotes the x-th learning principle in Graesser (2009).

principles cannot actually live without any teaching principle. For example, L02, L08, L09, L15, and L25 are all listed under the name of learning principles, but how the learning materials are presented (which more or less belongs to teaching principles) are involved by each of these five learning principles. More details can be found in Graesser (2009) if readers are interested in what these effects or principles refer to.

In addition, these principles are also strategies, though they are not organized in a hierarchical structure as Oxford (1990) or Dowson and McInerney (2004). However, it is almost certain that they have more concrete connotations in practice, because they are based on evidence, which means their effectiveness has been validated by empirical experiments. And empirical experiments require concrete and operational strategies. Our point here is not to build connections between the principles summarized by Graesser (2009) and the hierarchical structures proposed in Oxford (1990) or Dowson and McInerney (2004), but to remind readers that some strategies in the hierarchical structures have already been examined by empirical studies. And sooner or later, a huge but limited table full of learning strategies whose effectiveness is examined by empirical studies according to age, course, etc., will be finished, as in Table 5.2. And such table will bring us empirical parameters for learning effectiveness optimization, even if some parameters are inconsistent by different experiments.

CHAPTER 5

Applications in Vocabulary Learning

FIRST of all, the empirical principles (instructional principle 1 and learning principle 1, 5, and 16 in Table 4.2 in Chapter 4) tell us that there needs to be a time break to separate two processes of concept learning if necessary. Therefore, we can denote the blank interval for a break after learning concept c_i in space s_j as $\Delta t(i,j)$, then the time spent on learning T is

$$T = \sum_{i=1}^{n} \sum_{j=1}^{m} [t(i,j) + \Delta t(i,j)] \mathbf{I}(i,j), \tag{5.1}$$

where

$$\mathbf{I}(i,j) = \begin{cases} 1 & \text{learning } c_i \text{ in } s_j \\ 0 & \text{else} \end{cases}$$

Equation (5.1) is a catchup and also an extension of Equation (3.2) after empirical evidences are introduced and formally considered. However, this is not the solo reshaping.

5.1 PROBLEM STATEMENT

Generally speaking, if $t(i,j)$ increases, $\Delta t(i,j)$ will increase as well, since people may need more time for rest when they spend more time on brain consuming, e.g., learning (Hoover, Ragsdale, & Ayres, 2022; Cropley, Weidenstedt, Leick, & Sütterline, 2022; Hunter & Wu, 2016; Bakker, Demerouti, Oerlemans, & Sonnentag, 2013). Therefore, under

DOI: 10.4324/9781032707013-5

this circumstance, the solution on schema scheduling will not change, compared to Equation (2.2) in Chapter 2 where $\Delta t(i,j)$ is not taken into consideration at all.

However, $\Delta t(i,j)$ may not necessarily increase as $t(i,j)$ increases, though it is possible to increase as $t(i,j)$ increases. For example, when a learner p learns a simple concept c_i in a short time $t(i,j)$, if it believes such learning process is not worth its time compared to other things it wants to do (e.g., playing), it is more likely to cost more $\Delta t(i,j)$ than usual to adjust its mood or attention to learn the next concept. Such phenomenon exists both among seventh grade (Conley, 2012) and college students (Perez, Cromley, & Kaplan, 2014). On the contrary, it is also possible to decrease as $t(i,j)$ increases. When p learns a difficult concept c_l in a long time $t(l,j)$, if it believes that intelligence is a changeable trait that it can improve with effort and guidance (described as holding a "growth mindset" by Dweck (2006)), it is more likely to take on challenging tasks and persist through challenges by trying new strategies or increasing effort, ultimately achieving greater academic success (Smiley, Buttitta, Chung, Dubon, & Chang, 2016; Limeri et al., 2020; Yarborough & Fedesco, 2020). Therefore, it may not wait any longer $\Delta t(l,j)$ than usual to come to the next concept.

5.2 SOLUTION

5.2.1 Retention Rate

Retention rate (RR) refers to the ratio of the recalled content (after a certain period of blank time) recently learned over all the content recently learned. The English vocabulary retention rate can be retrieved or calculated by the evidences revealed in experimental studies on English vocabulary learning strategies. In the original Ebbinghaus' monograph (1885/1913), the forgetting curve described the changes of savings over time. Ebbinghaus defined savings as the relative amount of time saved on the second learning trial as a result of having had the first. However, after several decades, the often used forgetting functions turned to be based on recall probability (more details can be found in Murre and Chessa (2022)), which was essentially the same as the retention rate defined at the beginning of this subsection.

Since Ebbinghaus reported the classic forgetting curve (1885/1913), the form of the forgetting curve has been debated for over a century. Recently, Murre and Dros (2014) replicated Ebbinghaus' experiments successfully again and found that the good-of-fit of a simple power function to either their data or Ebbinghaus' data is quite impressive. A more complicated function named summed exponential function (sum of two exponentials) was yielded to fit the forgetting curve better (Murre, Chessa, & Metter, 2013). Such function was yielded from a model of forgetting and amnesia (Murre, Meeter, & Chessa, 2007), and therefore, it had a very solid explanation on its form.

However, the progress of Murre and Dros (2014) was based on Ebbinghaus' savings instead of the recall probability, and there is no simple transformation between these two indicators. Hence, we have to focus on the progress of recall-based fitting functions, so that the empirical data in vocabulary learning experiments, most of which in recent decades adopted recall probability to measure the effectiveness of learning strategies, can be used as much as possible. In recent three years, the studies on recall-based fitting include Plaskura (2019); Xie, Chen, and Li (2020); Peltokorpi and Jaber (2022). Plaskura (2019) compared the brain activities in the sphere of learning (knowledge learning and forgetting) to an electrical circuit network (e.g., battery charging and discharging) and designed three gradually refined circuits to model the learning process. By using a simulator to solve the differential equations of the electrical circuit networks, one of his circuit models can fit five of seven Ebbinghaus points. Xie et al. (2020) stepped forward; they assumed that the central executive of their modified Baddeley's memory model (1992) was linear time-invariant (an electrical circuit network is a typical linear time-invariant system). Although their assumption was debatable, their convolution forgetting curve model for repeated learning fitted the average forgetting data well. Besides these two studies whose authors had an educational background in information sciences, scholars who had a mechanical or industrial engineering background also contributed to the forgetting curve modeling. Peltokorpi and Jaber (2022) came up with a new idea; they incorporated production interruptions (or production breaks) into learning curve models. They presented an enhanced version of the power learning curve that accounted for a variable degree of interference when moving from a production cycle to the next. Such interference-adjusted model fitted the data very well. Their learning and forgetting model can help identify struggling workers early and release precocious trainers earlier than expected (Peltokorpi & Jaber, 2022).

Nevertheless, workers are not learners. Workers have strong economic interests in their jobs, so that they can keep similar enthusiasm on several production cycles, which brings sufficient empirical data of repeated learning and forgetting. Whereas some vocabulary learners may also have strong interests in word learning, such learners may only account for a small proportion of ordinary learners, which makes it difficult to cumulate enough empirical data to study the curves of repeated learning and forgetting, let alone many studies on multiple learning strategies. Therefore, we turned back to a commonly seen and well-established research scenario: a single session of learning and forgetting. That is, no matter how many strategies are involved, the learning process can be treated as a whole session, which will not appear repeatedly after the once-for-all learning; what can appear repeatedly is only the testing (how many words are still remembered along the time).

The single session of learning and forgetting is commonly seen in multiple summaries on vocabulary learning/teaching (Wegener, Beyersmann, Wang, & Castles, 2022; Jonathans, Widiati, Astutik, & Ratri, 2021; Rice & Tokowicz, 2020; Schmitt, 2019; Moody, Hu, Xu, & Lee, 2018). Besides these summaries on general vocabulary learning, there are also several reviews on computer-assisted vocabulary learning (Regina & Anitha, 2022; Zou, Huang, & Xie, 2021; J. Lin & Lin, 2019). Such summaries can be dated back to 2010. NRTAC (2010) filled NICHHD (2000)'s gap: there was little research that conclusively identified the best strategies or combinations of strategies of vocabulary instruction, and NRTAC (2010) collected several studies on the effectiveness of multiple vocabulary instruction strategies adopted by the same sample.

5.2.2 Asymptotic Retention Rate

As is well known, retention rate varies with the length of a break (also shown in Table 5.1). And even in the same study, the length of a break is not necessarily the same. Therefore, there is no widely followed standard of break length in the retention experiments. Hence, the empirical data have to be processed further, so that the comparisons among the effectiveness of vocabulary learning strategies can be laid on a fair base. Hereby, we decide to introduce the concept of asymptotic retention rate (ARR) used in Averell and Heathcote (2011).

Averell and Heathcote (2011) found that an exponential function with an asymptote as in Equation (5.2) provided the best fit to individual participant data collected under both explicit and implicit retrieval instructions.

$$R(t) = a + (1 - a) \times be^{-\alpha t} \tag{5.2}$$

$R(t)$ is the retention rate at the moment t. a is the asymptote parameter. Both a and b are assumed bounded between zero and one. Values of b less than one allow for the possibility that $R(0) < 1$ (Averell & Heathcote, 2011). α is a decaying parameter, which is larger than zero.

Figure 5.1 shows two exponential functions with asymptotes fitting two experimental datasets. These two datasets are provided in Nash and Snowling (2006) and listed in Table 5.1. The parameters of these two exponential functions are calculated when $R(0)$ is assumed to be 1. In fact, with this $R(0) = 1$ assumption, the parameters (a, b, and α) can be solved from Equation (5.2) when there are two extra data points, though two of the three equations are transcendental equations. By using the default algorithm ("trust-region-dogleg" (Powell, 1970)) built in fsolve function of MATLAB® 2022a, the numerical solutions of asymptotes (i.e., a) are obtained, as in Table 5.2. The initial values of all three parameters a, b and α are all set to 0.5 before the convergence starts.

5.2.3 Solving Asymptotic Retention Rate of a Single Learning Strategy by Linear Constraints

If we can assume that each vocabulary learning strategy contributes independently to the final asymptotic retention rate, we can model the final

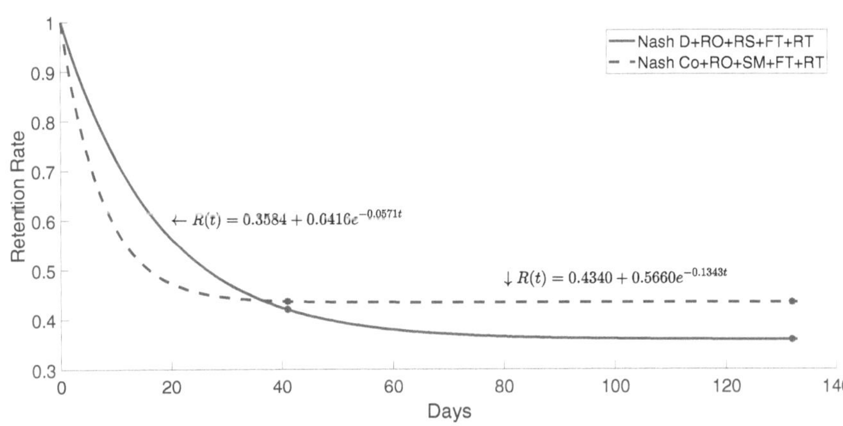

FIGURE 5.1 Two curves fitting two datasets in Nash and Snowling (2006).

TABLE 5.1 Vocabulary Instruction Strategies (or Their Combinations) and Their Retention Rates (Age 80–10)

Strategies Combination	Length of First Break	Retention Rate after First Break (%)	Length of Second Break	Retention Rate after Second Break (%)	Source
P	6 days	7.00	2 weeks	8.30	Jenkins, Pany, and Schreck (1978)
Co+S	3 days	16.67	2 weeks	9.67	Jenkins et al. (1978)
Co+D	3 days	44.50	2 weeks	22.17	Jenkins et al. (1978)
Ca+RO+FT+RT	1 session[a]	79.10	1 week	81.25	Nation, Snowling, and Clarke (2007)
S+Co+RO+FT+RT	3 days	83.33	2 weeks	23.67	Jenkins et al. (1978)
Co+RO+SM+FT+RT	6 weeks	43.63	3 months	43.40	Nash and Snowling (2006)
D+RO+RS+FT+RT	6 weeks	42.01	3 months	35.87	Nash and Snowling (2006)

D is for defining by dictionaries.
P is for presenting words.
S is for explaining with synonyms.
Ca is for using cards.
Co is for learning in a context.
FT is for feedback timely.
RO is for repeat orally.
RS is for repeat spelling.
RT is for recall timely.
SM is for using semantic maps.
[a] We have no further information from the source on how long a session is. Considering the ages of participants, we assume that one hour can be one of the best guesses.

asymptotic retention rate by a linear expression. That is,

$$ARR(S_i + S_j) = ARR(S_i) + ARR(S_j) \tag{5.3}$$

where ARR(·) is the asymptotic retention rate of ·, and S_i is the ith space for learning. In this subsection, $S_i \in S = \{$D, P, S, Ca, Co, FT, RO, RS, RT, SM$\}$ whose meanings are listed in the table notes of Table 5.1.

5.2.3.1 Using Linear Equations

According to Table 5.2, we have an equation array as follows:

$$\begin{bmatrix} 1 & 0 & 0 & 0 & 0 & 0 & 0 & 0 & 0 & 0 \\ 0 & 0 & 1 & 1 & 0 & 0 & 0 & 0 & 0 & 0 \\ 0 & 0 & 1 & 0 & 1 & 0 & 0 & 0 & 0 & 0 \\ 0 & 1 & 0 & 0 & 0 & 1 & 0 & 0 & 1 & 1 \\ 0 & 0 & 1 & 0 & 0 & 1 & 1 & 0 & 1 & 1 \\ 0 & 0 & 0 & 0 & 1 & 1 & 0 & 1 & 1 & 1 \end{bmatrix} \begin{bmatrix} ARR(P) \\ ARR(Ca) \\ ARR(Co) \\ ARR(S) \\ ARR(D) \\ ARR(RO) \\ ARR(SM) \\ ARR(RS) \\ ARR(FT) \\ ARR(RT) \end{bmatrix} = \begin{bmatrix} 7.65 \\ 9.67 \\ 22.16 \\ 80.18 \\ 43.40 \\ 35.84 \end{bmatrix} \tag{5.4}$$

The pseudo inverse of the left array (6 × 10) in Equation (5.4) is the solution of Equation (5.4), and the 2-norm of this solution is the minimum. That is,

$$\begin{bmatrix} ARR(P) \\ ARR(Ca) \\ ARR(Co) \\ ARR(S) \\ ARR(D) \\ ARR(RO) \\ ARR(SM) \\ ARR(RS) \\ ARR(FT) \\ ARR(RT) \end{bmatrix} = \begin{bmatrix} 7.650 \\ 39.002 \\ 12.310 \\ -2.640 \\ 9.850 \\ 13.726 \\ -10.088 \\ -15.188 \\ 13.726 \\ 13.726 \end{bmatrix} \tag{5.5}$$

The negative ARRs seem to be problematic. However, it may also imply that some strategies may undermine others' performance when they are combined together. For example, those strategies whose ARR signs are negative, e.g., "S" or "SM", may interfere with other strategies, such as "Ca". And indeed, we have found the empirical evidences!

Early in 1960s, Higa (1963) found that among the four association word lists, only the lists of strong associates and antonyms were more difficult than the control word list (no-association). Of the two semantic distance word lists, only the synonym word list was more difficult than the control word list. Learning the control word list first facilitated the learning of an experimental word list significantly more than vice versa. And a growing body of empirical evidences that suggest semantic clustering deters L2 vocabulary acquisition at the novice level (Wilcox & Medina, 2013; Tagashira, Kida, & Hoshino, 2010; Erten & Tekin,

TABLE 5.2 Retention Rates (RR) and Asymptotic Retention Rates (ARR) of Vocabulary Instruction Strategies (Age 8–10) in Percents[a]

Strategies Combination	ARR	First RR	Second RR	Source
P	7.65	7.00	8.30	Jenkins et al. (1978)
Co+S	9.67	16.67	9.67	Jenkins et al. (1978)
Co+D	22.16	44.50	22.17	Jenkins et al. (1978)
Ca+RO+FT+RT	80.18	79.10	81.25	Nation et al. (2007)
S+Co+RO+FT+RT	0.00 [b]	83.33	23.67	Jenkins et al. (1978)
Co+RO+SM+FT+RT	43.40	43.63	43.40	Nash and Snowling (2006)
D+RO+RS+FT+RT	35.84	42.01	35.87	Nash and Snowling (2006)

[a] The abbreviations are the same as the table notes of Table 5.1.

[b] The calculated ARR is −0.0513 (−5.13%). We assume that 0.00 will make more sense. Jenkins et al. (1978) did not analyze the reason for the dramatic drop from first RR to second RR, though the combination of the learning strategies in this row delivered the best RR among the four combinations in Jenkins et al. (1978). As far as we know, the order of the learning strategies in such combination may effect the RRs, since it is the only combination that begins with S (explaining with Synonyms). In our opinion, although it is the highest RR among the four RRs in Jenkins et al. (1978), the dramatic drop shows its vulnerability to some extent. And considering its calculated ARR is negative, we have decided to excludethis data in our coming explorations.

2008; Finkbeiner & Nicol, 2003; Waring, 1997; Tinkham, 1997, 1993). Specifically, learning semantically related words took about 50% longer than learning semantically unrelated words (Tagashira et al., 2010).

Semantically clustered word groupings are more difficult to learn than non-related groups of words among novice learners (Wilcox & Medina, 2013). What's worse, this difficulty in the short term did not translate into better long-term retention (Schneider, Healy, & Bourne, 2002). Interested readers may go to the "theoretical framework" subsection in Wilcox and Medina (2013) for a deep explanation. In short, the mind may initially be prepared to receive words that need to be organized in semantic fields, not to receive words that have already been semantically pre-organized (Wilcox & Medina, 2013).

Therefore, it seems rather reasonable to have negative ARRs when we discuss the effectiveness of a single learning strategy in the context of multiple strategies working together.

5.2.3.2 Using Optimization Tools

To avoid negative ARRs when we pursuit the effectiveness of a single learning strategy without the context of strategy combinations, and also considering that a minimum 2-norm has its appropriate meaning in our scenario (which is, all the ARRs ought to be small enough since RR will decrease as time goes by), we turn (5.4) to an optimization problem, which is also a nonlinear programming problem:

$$\begin{aligned} &\min \sum_{S_i \in S} ARR(S_i)^2 \\ &s.t. \quad \begin{cases} \text{Aeq} * \text{ARR} = \text{beq} \\ \forall S_i, ARR(S_i) \geq 0 \end{cases} \end{aligned} \tag{5.6}$$

where

$$ARR = \begin{bmatrix} ARR(P) \\ ARR(Ca) \\ ARR(Co) \\ ARR(S) \\ ARR(D) \\ ARR(RO) \\ ARR(SM) \\ ARR(RS) \\ ARR(FT) \\ ARR(RT) \end{bmatrix} \tag{5.7}$$

and,

$$\text{Aeq} = \begin{bmatrix} 1 & 0 & 0 & 0 & 0 & 0 & 0 & 0 & 0 & 0 \\ 0 & 0 & 1 & 1 & 0 & 0 & 0 & 0 & 0 & 0 \\ 0 & 0 & 1 & 0 & 1 & 0 & 0 & 0 & 0 & 0 \\ 0 & 1 & 0 & 0 & 0 & 1 & 0 & 0 & 1 & 1 \\ 0 & 0 & 1 & 0 & 0 & 1 & 1 & 0 & 1 & 1 \\ 0 & 0 & 0 & 0 & 1 & 1 & 0 & 1 & 1 & 1 \end{bmatrix} \quad \text{beq} = \begin{bmatrix} 7.65 \\ 9.67 \\ 22.16 \\ 80.18 \\ 43.40 \\ 35.84 \end{bmatrix}$$

Then, by using the default algorithm ("interior-point" (Waltz, Morales, Nocedal, & Orban, 2006)) built in the fmincon function of MATLAB®

2022a, we have the optimal solution ARR^*:

$$ARR^* = \begin{bmatrix} 7.65 \\ 56.83 \\ 9.67 \\ 0.00 \\ 12.49 \\ 7.78 \\ 10.38 \\ 0.00 \\ 7.78 \\ 7.78 \end{bmatrix} \tag{5.8}$$

If we change the objective function (2-norm) in Equation (5.6) into a 1-norm form, which is $\min \sum_{S_i \in S} ARR(S_i)$, it can also reflect our expectation that all the ARRs ought to be small enough, though it is now a linear programming problem. Under the same constraints, we have the optimal solution ARR$^{'}$, by using the default algorithm ("dual-simplex" (Dantzig, Orden, & Wolfe, 1955)) built in the linprog function of MATLAB® 2022a:

$$ARR^{'} = \begin{bmatrix} 7.65 \\ 56.83 \\ 9.67 \\ 0.00 \\ 12.49 \\ 0.00 \\ 10.38 \\ 0.00 \\ 0.00 \\ 23.35 \end{bmatrix} \tag{5.9}$$

Apparently, ARR* $\neq$ $ARR^{'}$, but we have $\sum_{S_i \in S} ARR^*(S_i) = \sum_{S_i \in S} ARR'(S_i)$. In the following theorem, we can prove that not only in this specific case but also in a much broader range that a 2-norm minimum will lead to a 1-norm minimum.

Theorem 5.1. *$\forall x_i \geq 0 (i = 1, 2, \ldots, n)$, if X^*_{2-norm} is an optimal solution to $\min f(x_1, x_2, \ldots, x_n) = \sum_{i=1}^{n} x_i^2$, with a set of constraints C, then X^*_{2-norm} is also an optimal solution to $\min g(x_1, x_2, \ldots, x_n) = \sum_{i=1}^{n} x_i$ with the same C.*

Proof. If $\forall x_i \in \mathbb{R}$, then the solution to $\sum_{i=1}^{n} \frac{\partial f}{\partial x_i} = 0$ is the optimal solution to $\min f(x_1, x_2, \ldots, x_n)$. However, $\forall x_i \geq 0 (i = 1, 2, \ldots, n)$ and with the set of constraints C, $\sum_{i=1}^{n} \frac{\partial f}{\partial x_i}$ may not always reach 0.

$\because f(x_1, x_2, \ldots, x_n)$ is a monotonically increasing function of $x_i, i = 1, 2, \ldots, n)$ and $\forall x_i \geq 0 (i = 1, 2, \ldots, n)$, $\frac{\partial^2 f}{\partial x_i^2} > 0$,

$\therefore \sum_{i=1}^{n} \frac{\partial f}{\partial x_i}$ has to be the minimum, to obtain the optimal solution to $\min f(x_1, x_2, \ldots, x_n)$.

$\because \sum_{i=1}^{n} \frac{\partial f}{\partial x_i} = 2 \sum_{i=1}^{n} x_i = 2g(x_1, x_2, \ldots, x_n)$

$\therefore g(x_1, x_2, \ldots, x_n)$ reaches its minimum with the same C, too, when $\sum_{i=1}^{n} \frac{\partial f}{\partial x_i}$ reaches the minimum.

□

Therefore, ARR* is also one of optimal solutions to $\min \sum_{S_i \in S} ARR(S_i)$ with the same constraints in Equation (5.6). The computer programs may only return one of multiple optimal solutions.

Under the constraints of non-negativity, there is not a negative result in the optimal solution, which is much better than the negative-ARR-solution. However, both of these two non-negative solutions still contain zeros. Although in the long run, we forget all the knowledge we learn, ("In the long run, we are all dead", John Maynard Keyns), we expect to pursuit a full positive ARR solution since it can indicate the effectiveness of a learning strategy or a space for learning.

5.2.3.3 Back to the Start: Spiraling Up

We need to look backward at the original (5.4) and come up with a new idea. Before we present this new idea, we would like to introduce the "mean vertex" on a convex polyhedra. Figure 5.2 demonstrates a convex "polyhedra"(polygon, actually) bounded by a set of constraints: $x+y+z = 1, x \geq 0, y \geq 0$, and $z \geq 0$. The vertices are (1,0,0), (0,1,0), and (0,0,1). The mean vertex is a point whose coordinates are the sum of all the vertices of a convex polyhedra divided by the number of vertices. In Figure 5.2, the mean vertex is $[(1, 0, 0) + (0, 1, 0) + (0, 0, 1)]/3 = (1/3, 1/3, 1/3)$. Please note that the mean vertex is one of the solutions to the set of constraints.[1] The mean vertex (i.e., centroid) of a convex polyhedra can be a representative of the elements in the scope of convex polyhedra, as in Shaffer et al. (2016).

[1] The proof is simple. The mean vertex is actually the centroid of a convex polyhedra. If the centroid of a convex polyhedra is not in the convex polyhedra, the polyhedra is not convex.

So, the mean vertex solution is also a representative solution that satisfies the set of constraints.

Using the programs contributed by Matt (2021), the mean vertex solution to Equation (5.4) is:

$$ARR_{\text{mean\_vertex}} = \begin{bmatrix} 7.650 \\ 66.294 \\ 4.835 \\ 4.835 \\ 17.325 \\ 4.629 \\ 24.679 \\ 4.629 \\ 4.629 \\ 4.629 \end{bmatrix} \tag{5.10}$$

This time, we obtain all the positive and representative ARRs. Among the ten learning strategies in the set S, the largest ARR (or the most effective vocabulary learning strategies in the terms of ARR) belongs to S_2, whose abbreviation is Ca (i.e., Card). And such mathematical result is consistent with people's daily experiences. Even nowadays, Chinese pupils are often using Card-like cell-phone applications to learn English words (Ma, 2022). Two rigorous academic papers even adopted card (J. Lin & Lin, 2019) or

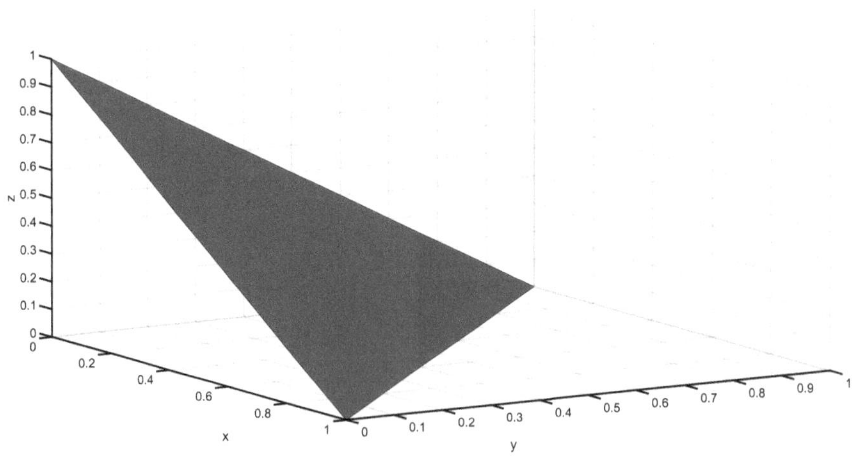

FIGURE 5.2 An example of convex polygon.

online flashcard (Hao, Wang, & Ardasheva, 2021) as a potential benchmark to meta-analyze the effectiveness of mobile apps on vocabulary learning, which implies that card is indeed an effective learning strategy or probably the most effective strategy among traditional and single spaces for learning.

A keen reader may argue the assumption (5.3) does not hold: if we add up all the elements in $\text{ARR}_{\text{mean\_vertex}}$, the sum will be larger than 100. This means there exists a combination of vocabulary learning strategies whose asymptotic retention rate is larger than 100%; however, this will not happen in the physical world. How about adding an additional constraint (i.e., $\sum_{S_i \in S} ARR(S_i) \leq 100$) into (5.4)? Unfortunately, the program of Matt (2021) falsifies that the new constraints can form a convex polyhedra. Together with the implications from Equation (5.5), which are presented below Equation (5.5), we believe that ARR$(S_i + S_j) \leq ARR(S_i) + ARR(S_j)$ will be more close to the physical world. In the future, we will examine if this updated constraint can bring us more surprises.

5.2.3.4 Connections to Matroid Theory: Submodular

If S_i (i = 1,2,…,n) does not have any interaction with each other, or S_i indicates the atomic-level vocabulary learning strategy that cannot be divided any further, which is also entailed in the former assumption that "each vocabulary learning strategy contributes independently to the final asymptotic retention rate", we have ARR$(S_i \cap S_j) = 0$ since $S_i \cap S_j = \emptyset$. And then, we have ARR$(S_i + S_j) + ARR(S_i \cap S_j) \leq ARR(S_i) + ARR(S_j)$, or ARR$(S_i \cup S_j) + ARR(S_i \cap S_j) \leq ARR(S_i) + ARR(S_j)$. In matroid theory, this inequality informs us that ARR is a submodular function. It means that as the number of combined strategies increases, the increased learning effectiveness (hereby ARR) decreases.

In fact, though Table 5.2 is short, it shows a symptom already: if we can treat the combination of RO, FT, and RT as a subset, one strategy (Ca) can increase ARR of this subset to 80.18% (Row 4); on the contrary, two strategies (Co ∪ SM) fail to increase ARR of this subset over 80.18% (Row 6, only 43.40%, actually), let alone the "increased" ARR. On the contrary, if we alter the perspective (i.e., we are not going to add one or two strategies into the subset {RO, FT, RT}, we add {RO, FT, RT} into {Ca} or {Co, SM} instead), it is reasonable to expect that the incremental value of adding {RO, FT, RT} into a smaller set ({Ca}) is greater than adding it into a larger set ({Co, SM}), though we do not have the empirical evidence.

But in theory, it is proved that if a function is submodular, the incremental value of adding new elements into a smaller set will be greater than adding them into a larger set (i.e., the law of diminishing marginal utility (Gossen, 1854)). We present the proof extended from Narayanan (1997, p. 327) as follows.

Theorem 5.2. *Let S be a finite set. Let* $f: 2^S \rightarrow \mathbb{R}$, *and f is a submodular function, i.e.,* $\forall X, Y \subseteq S$: $f(X) + f(Y) \geq f(X \cup Y) + f(X \cap Y)$. *Then,* $f(X \cup a) - f(X) \geq f(X \cup b \cup a) - f(X \cup b), \forall X \subseteq S, \forall a, b \subseteq S \setminus X$.

Proof 5.2. $\because f(X) + f(Y) \geq f(X \cup Y) + f(X \cap Y)$,
Let $X \leftarrow X \cup a$, and $Y \leftarrow X \cup b$.
$\therefore f(X \cup a) + f(X \cup b) \geq f((X \cup a) \cup (X \cup b)) + f((X \cup a) \cap (X \cup b))$,
$\therefore f(X \cup a) + f(X \cup b) \geq f(X \cup b \cup a) + f(X)$,
$\therefore f(X \cup a) - f(X) \geq f(X \cup b \cup a) - f(X \cup b)$

□

Therefore, the incremental value of adding the set a into a smaller set X will be greater than adding it into a larger set $X \cup b$.

Figure 5.3 shows that the relationship between the number of strategies combined and ARR can be fit well by two curves, namely, Gauss model and polynomial model. Either model shows that the mapping function of vocabulary learning strategies and their effectiveness (ARR) is a non-monotone submodular.

The sum of squares for error (SSE) of the Gauss model is 142.8, and the number of parameters of Gauss model is 3. And SSE of the polynomial model is 106.6, and the number of parameters of polynomial model is 4. According to Akaike (1974)'s information criterion (AIC) in the least square case, AIC$= n\ln(\mathrm{SSE}/n)+2K$ (Burnham & Anderson, 2002, p. 63), where K is the number of parameters ($K = 3$ in the Gauss model, and $K = 4$ in the polynomial model), n is the number of samples ($n = 5$ in both models). Then, the Gauss model's AIC is 22.76 and the polynomial model's AIC is 23.30. Under AIC, a model with a smaller AIC value is better. Therefore, the Gauss model is better.

However, Gauss model and polynomial model are both non-monotone, and the progresses on maximizing monotone submodular function (e.g., (Qian, Yu, Tang, Yao, & Zhou, 2019)) are more advanced than the ones on maximizing non-monotone submodular function (e.g., (Feige, Mirrokni, & Vondrak, 2011)). Therefore, though we can observe the Gauss curve by eye

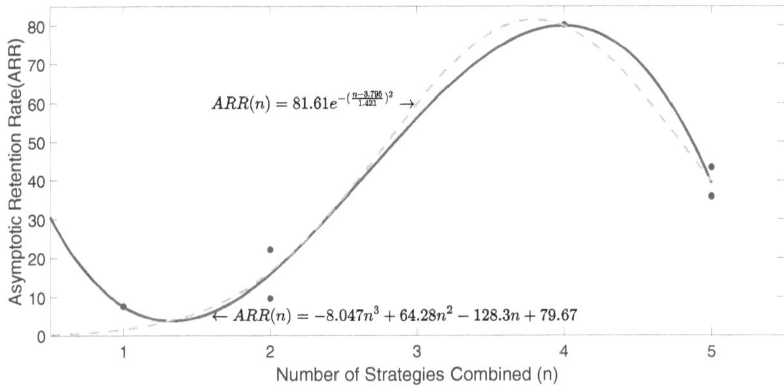

FIGURE 5.3 Two curves fitting the relationship between the number of strategies combined and asymptotic retention rate.

to obtain the solution to the optimization problem, we will investigate on a generalized and abstract approach to solve this kind of problems in the future.

5.2.4 Predicting Asymptotic Retention Rate by Neural Network

In the previous subsection, we can solve the asymptotic retention rate of a single learning strategy by linear constraints. The original set of constraints in Equation (5.4) indicates that different learning strategies may interfere with each other when they are used in a same combination, in theory. Following this lead, we find several empirical evidences reporting that different learning strategies interfere with each other in vocabulary learning. The refined set of constraints in Equation (5.6) and a representative solution of Equation (5.4) as in (5.10) both indicate that "card" is the most effective strategy (in terms of asymptotic retention rate) among the involved vocabulary learning strategies when being used alone, in theory. And again in practice, we find that several review articles use "card" as one of the benchmarks to compare its effectiveness with the ones of other mixed vocabulary learning strategies, which implies that it is the most effective single strategy according to scholars' studies or students' daily experiences.

In this subsection, we are going to demonstrate whether the asymptotic retention rate of a strategy combination or a mixed space for learning can be predicted, by the already existed evidences listed in Table 5.2.

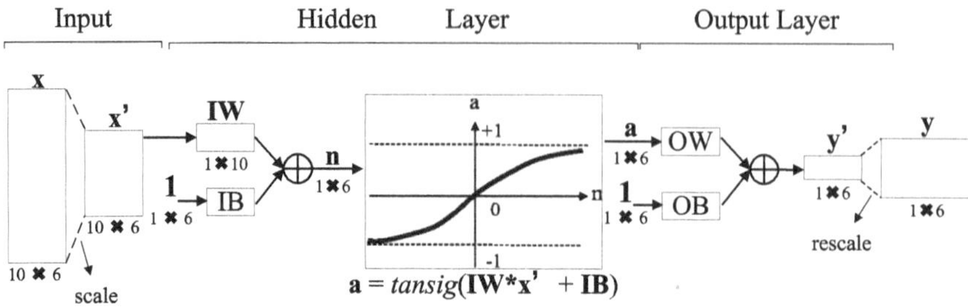

FIGURE 5.4 The architecture of a forward neural network.

Asymptotic retention rate can be regarded as a function of word learning strategies. Kreinovich (1991) and other scholars proved that any continuous function can be approximated at any precision by neural networks. Such neural network usually consists of many linear neurons and a neuron implementing an arbitrary but smooth nonlinear function. However, neural network does not necessarily learn representations that can be easily interpreted by human beings. And how to actively make a network interpretable without harming its performance is still an open problem (Zhang, Tiňo, Leonardis, & Tang, 2021). Therefore, we decide to keep the architecture of neural network as simple as possible: a forward neural network with only two linear neurons and one nonlinear neuron. Its architecture and its detailed calculating flow chart are presented in Figures 5.4 and 5.5, respectively.

The architecture and flow chart tell people how a neural network connects input data (i.e., strategy combination) to output data (i.e., asymptotic retention rate) and leave the weights and biases of neurons to be decided. IW_1, IW_2,. . ., IW_{10} and IB are input weights and input bias. OW and OB are output weight and output bias. The weights and biases of an neural network are originally generated by a random seed, and then refined step by step, until the ending condition of iterations is satisfied. The specific method is the Levenberg-Marquardt algorithm. More details can be found at Chapters 11 and 12 in Hagan, Demuth, Beale, and De Jesús (2014).

It seems that it is difficult to obtain a neural network with just two linear neurons whose accuracy can be acceptable, especially when there are only six samples[2] (Six ARRs derived from experiments in literatures in our

[2] Four samples are randomly chosen as training samples while the left two are for test. And during the training, the specific training samples will experience multiple replacements, only the number of training samples and the number of test samples are fixed to four and two, respectively.

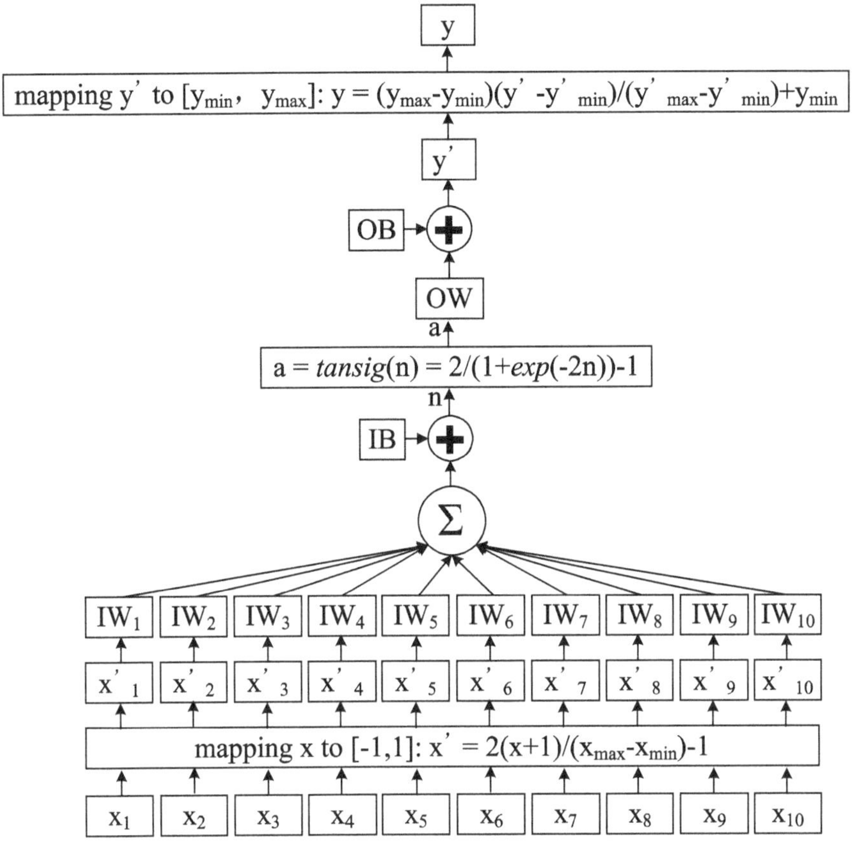

FIGURE 5.5 The detailed calculating flow of the forward neural network.

study, i.e., the right vector in Equation (5.4)) and ten features (ten single or combined word learning strategies derived from experiments in literatures in our study, i.e., the left matrix in Equation (5.4)). However, we trained out two acceptable sets of parameters for a forward neural network within just half an hour after its first official run by a common personal computer (Intel® Core™ i7-8700 CPU @ 3.20 GHz, 3.19 GHz), indicating that the connections behind strategies and ARRs are much stronger than it seems. Their values are listed in Table 5.3.

Table 5.3 lists two sets of parameters for the neural network presented in Figures 5.4 and 5.5. The parameters not only include the basic parameters for a neural network, such as weights and biases, but also include the result:

TABLE 5.3 Two Sets of Parameters for Neural Network

	IW_1	**IW_2**	**IW_3**	**IW_4**	**IW_5**	**IW_6**	**IW_7**	**IW_8**	**IW_9**	**IW_{10}**
First	0.4970	−0.9067	0.0837	0.1671	−0.4315	0.3616	−0.3254	0.2956	−0.0969	−0.7653
Second	−0.7547	−1.3200	1.3524	−0.4793	1.7781	−1.2968	−0.2058	−0.5364	−0.5768	−0.6690
	IB	**OW**	**OB**	y_1	y_2	y_3	y_4	y_5	y_6	**M.S.E.**
First	0.1413	−0.8006	−0.1894	8.64	9.67	22.16	63.59	43.40	37.36	46.43
Second	.3584	−0.8000	0.2001	24.65	22.16	22.16	80.18	43.40	38.58	75.30

TABLE 5.4 Performances (M.S.E.) of First Six Trials vs. Sizes of Hidden Layer

No. of Hidden Neurons	Trial 1	Trial 2	Trial 3	Trial 4	Trial 5	Trial 6	Sum
1	165.29	776.71	492.48	303.00	762.52	1631.4	4131.4
2	927.93	3524.8	256.02	1008.5	367.31	490.22	6574.8
3	1273.6	1618.1	922.57	840.23	1150.6	309.32	6114.4
4	726.28	737.33	3119.5	1510.5	566.80	145.56	6806.0
5	1011.4	452.42	352.79	998.31	1902.5	759.33	5476.8

predicted ARRs (i.e., $y_i, i = 1, 2, 3, \ldots, 6$) and a reference for performance evaluation: M.S.E.(mean squared error between the predicted ARRs and the observed ARRs, i.e., the right vector in Equation (5.4)). In terms of M.S.E., the corresponding neural network with the 1st set of parameters outperforms the one with the second set of parameters. In terms of a single ARR, though the second set of parameters can make a neural network predict the highest ARR (80.18) very precisely, it fails on other low ARRs, especially on y_1 and y_2. Overall, the first set of parameters delivers the best forward neural network.

5.2.4.1 Connections to Matroid Theory: Greedy Algorithm

Lee, Bartlett, and Williamson (1996) proved that it is possible to approximate functions in the class of bounded fan-in neural networks, whose sum of absolute values of output weights is also bounded, accurately with relatively few hidden units. The approach is using greedy algorithm: iteratively adding hidden units greedily to reduce the approximation error.

Recently, Paria, Poczos, Ravikumar, Schneider, and Suggala (2022) found that to solve the black-box optimization, a simple greedy algorithm that fits a neural network to the current set of observations and uses the learned network as the acquisition function to determine the next query point often outperforms other acquiring methods, such as Thompson Sampling, Upper Confidence Bound, and Expected Improvement. In the previous section, we started with one hidden unit, and it has already shown an acceptable accuracy. In fact, in this case, more hidden units do not help or make the performance even worse, as seen in Table 5.4. It shows the performances of first six trials when the number of hidden units is set from one to five (the number of training samples and the number of test samples are still four and two, respectively). So, we stop at one hidden unit by the greedy algorithm.

CHAPTER 6

Applications in Math Learning

SUPPOSE in a mathematical course, there are N concepts, say C_i, $i = 1, 2, \ldots, N$, and there are M learning strategies or spaces for learning, denoted by S_k, $k = 1, 2, \ldots, M$. When one learns concept C_i in a space for learning or with a learning strategy S_k, the learning duration (time length) is $L(C_i, S_k)$, starting from the moment $L_s(C_i, S_k)$, and ending at the moment $L_e(C_i, S_k)$, i.e., $L(C_i, S_k) = L_e(C_i, S_k) - L_s(C_i, S_k)$.

After the moment $L_e(C_i, S_k)$, the learner has to rest from the moment $R_s(C_i)$ to the moment $R_e(C_i)$. The rest duration is $R(C_i) = R_e(C_i) - R_s(C_i)$. And $R_s(C_i) = L_e(C_i, S_k)$.

The effective duration of adopting strategy S_k or being in a space for learning, S_k, to learn C_i, is $E(C_i, S_k)$, starting from the moment $E_s(C_i, S_k)$ and ending at the moment $E_e(C_i, S_k)$, i.e., $E(C_i, S_k) = E_e(C_i, S_k) - E_s(C_i, S_k)$. And $E_s(C_i, S_k) = L_e(C_i, S_k) = R_s(C_i)$.

During the effective interval $[E_s(C_i, S_k), E_e(C_i, S_k)]$, concept C_i is assumed to be mastered by the learner 100% perfectly well, while the effective interval ends at the moment $E_e(C_i, S_k)$, the one is assumed to not master concept C_i at all, which leads all the descendant concepts to be forgotten.

How do we schedule the learning strategies so that the concepts can be retained as long as possible, given that learning each concept must be separated by a blank time for rest?

DOI: 10.4324/9781032707013-6

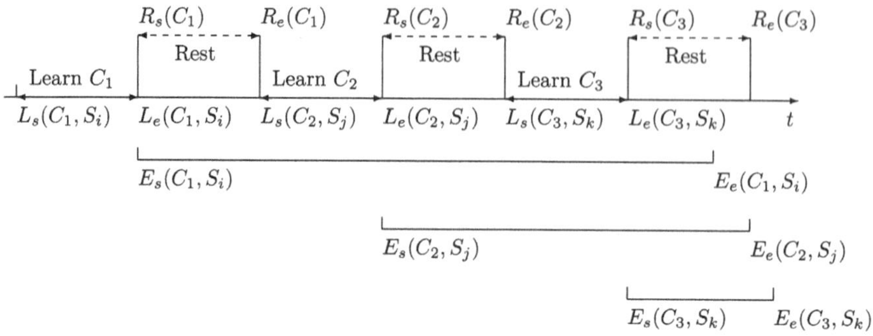

FIGURE 6.1 An example of scheduling (ideal).

Figure 6.1 illustrates learning, rest, and effective intervals of three concepts and shows an example of the notations above. In this example, a learner learns C_1, C_2, and C_3 with S_i, S_j, and S_k, and each learning is followed by a rest.

If there is only one concept (C_1) in this mathematical course, the maximum effective duration is:

$$T_{1\_\text{max}} = E_e(C_1, S_i) - E_s(C_1, S_i) \tag{6.1}$$

If there are two concepts (C_1, C_2) in this mathematical course, the maximum effective duration is:

$$T_{2\_\text{max}} = E_e(C_1, S_i) - E_s(C_2, S_j) \tag{6.2}$$

This equation holds under the constraint:

$$E_e(C_2, S_j) \geq E_e(C_1, S_i) \tag{6.3}$$

If there are three concepts (C_1, C_2) in this mathematical course, the maximum effective duration is:

$$T_{3\_\text{max}} = E_e(C_1, S_i) - E_s(C_3, S_k) \tag{6.4}$$

This equation holds under the constraints (6.3) and (6.5):

$$E_e(C_3, S_k) \geq E_e(C_2, S_j) \tag{6.5}$$

In fact, Figure 6.1 is an ideal scheduling that each invalid moment of a concept (e.g., $E_e(C_3, S_k)$) is later than the one of its precedent concepts (e.g., $E_e(C_2, S_j)$).

6.1 IDEAL SCHEDULING

According to Equations (6.2) and (6.1),

$$\begin{aligned} T_{2\_\max} &= E_e(C_1, S_i) - E_s(C_1, S_i) + E_s(C_1, S_i) - E_s(C_2, S_j) \\ &= T_{1\_\max} - [E_s(C_2, S_j) - E_s(C_1, S_i)] \end{aligned} \tag{6.6}$$

Similarly,

$$T_{3\_\max} = T_{2\_\max} - [E_s(C_3, S_k) - E_s(C_2, S_j)] \tag{6.7}$$

From Equation (6.6), we can conclude that, if $T_{1\_\max}$ has been fixed, $T_{2\_\max}$ will be obtained if and only if $[E_s(C_2, S_j) - E_s(C_1, S_i)]$ reaches the minimum value. Similarly, according to Equation (6.7), if $T_{2\_\max}$ has been fixed, $T_{3\_\max}$ will be obtained if and only if $[E_s(C_3, S_k) - E_s(C_2, S_j)]$ reaches the minimum value. Now we look into $[E_s(C_3, S_k) - E_s(C_2, S_j)]$ (similar conclusion can also apply to $[E_s(C_2, S_j) - E_s(C_1, S_i)]$). Since $[E_s(C_3, S_k) - E_s(C_2, S_j)] = R(C_2) + L(C_3, S_k)$ and $R(C_2)$ is usually constant, $[E_s(C_3, S_k) - E_s(C_2, S_j)]$ will reach the minimum value if and only if $L(C_3, S_k)$ reaches the minimum value. So, this observation tells us that if a learner wants to extend $T_{3\_\max}(C_3, S_k)$ as long as possible, this learner may choose an S_k whose learning duration is the minimum when $T_{2\_\max}$ has already been obtained.

On the other hand, by constraint (6.5), $E_e(C_3, S_k)$ should be as late as possible; by Equation (6.4), $E_s(C_3, S_k)$ should be as early as possible. Therefore, $E(C_3, S_k) = E_e(C_3, S_k) - E_s(C_3, S_k)$ should be as long as possible. Together with the implication observed in the previous paragraph, for a single concept, we should greedily choose a strategy whose effective duration is long enough, and also learning duration is short enough, so that the final intersecting effective duration ($T_{3\_\max}$ in Figure 6.1) covering all the concepts can be the longest. However, how do we make a choice when there is no strategy whose effective duration is the longest and the learning duration is the shortest?

6.2 LONGEST EFFECTIVE DURATION VS. SHORTEST LEARNING DURATION

6.2.1 A Numeric Example

Suppose we have a two-concept mathematical course (C_1 and C_2, and C_1 is the precedent concept of C_2) and each concept has two learning strategies

TABLE 6.1 Parameters

	$k = 1$	$k = 2$
$L(\cdot, S_k)$	1	2
$E(\cdot, S_k)$	5	8
$R(\cdot)$	0.1	0.1

(S_1, S_2) to choose. In this very simple example, learning strategies are assumed to be independent from concepts, which means neither $L(C_i, S_k)$ nor $E(C_i, S_k)$ $(k = 1, 2)$ is affected by C_i $(i = 1, 2)$. The specific values of $L(\cdot, S_k)$, $E(\cdot, S_k)$ and $R(C_i)$ are listed in Table 6.1.

Using these parameters, we can calculate $T_{1\_\max}$ and $T_{2\_\max}$ in two learning phases, respectively. According to the precedence relationship between C_1 and C_2, the first learning phase is $< C_1, S_1/S_2 >$ (indicating the learner adopts strategy S_1 or S_2 to learn C_1), and the second one is $< C_2, S_1/S_2 >$. We denote the whole learning procedure by $< C_1, S_1/S_2 > \to < C_2, S_1/S_2 >$.

First phase comes first. In $< C_1, S_1 >$, let $L_s(C_1, S_1) = 0$, then $L_e(C_1, S_1) = L_s(C_1, S_1) + L(C_1, S_1) = 0 + 1 = 1$, and $T_{1\_\max}(C_1, S_1) = 5$. Similarly, in $< C_1, S_2 >$, $T_{1\_\max}(C_1, S_2) = 8$. In this phase (or a course which has only one concept), it is obviously that in terms of effective duration, the strategy with longer effective duration outperforms the one with shorter effective duration. In the second phase, we divide this phase into four cases and discuss them separately.

- Case 1: $< C_1, S_1 > \to < C_2, S_1 >$

When the second phase joins in our consideration, the situation becomes complicated. In the learning procedure $< C_1, S_1 > \to < C_2, S_1 >$, let $L_s(C_1, S_1) = 0$, then we have the following equations.

$$\begin{aligned}
L_e(C_1, S_1) &= 1 \\
E_e(C_1, S_1) &= E_s(C_1, S_1) + E(C_1, S_1) = L_e(C_1, S_1) + E(C_1, S_1) \\
&= 1 + 5 = 6 \\
L_s(C_2, S_1) &= L_e(C_1, S_1) + R(C_1) = 1 + 0.1 = 1.1 \\
E_s(C_2, S_1) &= L_e(C_2, S_1) = L_s(C_2, S_1) + L(C_2, S_1) = 1.1 + 1 = 2.1 \\
E_e(C_2, S_1) &= E_s(C_2, S_1) + E(C_2, S_1) = 2.1 + 5 = 7.1
\end{aligned}$$

Since $E_e(C_2, S_1) > E_e(C_1, S_1)$, the constraint (6.3) is satisfied.

Therefore, $T_{2\_max} = E_e(C_1, S_1) - E_s(C_2, S_1) = 6 - 2.1 = 3.9$. It also passes the double check when using Equation (6.6). In the coming cases, for brevity we will not mention the results pass the double check by Equation (6.6) though we have done so.

- Case 2: $< C_1, S_1 > \rightarrow < C_2, S_2 >$

Let $L_s(C_1, S_1) = 0$, then, we have the following equations.

$$\begin{aligned}
L_e(C_1, S_1) &= 1 \\
E_e(C_1, S_1) &= E_s(C_1, S_1) + E(C_1, S_1) = L_e(C_1, S_1) + E(C_1, S_1) \\
&= 1 + 5 = 6 \\
L_s(C_2, S_2) &= L_e(C_1, S_1) + R(C_1) = 1 + 0.1 = 1.1 \\
E_s(C_2, S_2) &= L_e(C_2, S_2) = L_s(C_2, S_2) + L(C_2, S_2) = 1.1 + 2 = 3.1 \\
E_e(C_2, S_2) &= E_s(C_2, S_2) + E(C_2, S_2) = 3.1 + 8 = 11.1
\end{aligned}$$

Since $E_e(C_2, S_2) > E_e(C_1, S_1)$, the constraint (6.3) is satisfied.

Therefore, $T_{2\_max} = E_e(C_1, S_1) - E_s(C_2, S_2) = 6 - 3.1 = 2.9$. Comparing it to the result in the previous case, it is reasonable to speculate that **Speculation (1)** if T_{1_max} is the same, in terms of T_{2_max}, the strategy (i.e., S_1) applied to the next concept C_2 with a shorter learning duration outperforms the one (i.e., S_2) with longer learning duration. Under this circumstance, the effective durations of strategies do not matter. And in Table 6.1, it is dedicated to set a shorter learning duration combined with a shorter effective duration (S_1). Therefore, **Speculation (1)** can be rephrased in an anti-intuitive approach: if T_{1_max} is the same, in terms of T_{2_max} the strategy (i.e., S_1) applied to the next concept C_2 with a shorter effective duration outperforms the one (i.e., S_2) with longer effective duration.

- Case 3: $< C_1, S_2 > \rightarrow < C_2, S_1 >$

Let $L_s(C_1, S_2) = 0$, then we have the following equations.

$$\begin{aligned}
L_e(C_1, S_2) &= 2 \\
E_e(C_1, S_2) &= E_s(C_1, S_2) + E(C_1, S_2) = L_e(C_1, S_2) + E(C_1, S_2) \\
&= 2 + 8 = 10 \\
L_s(C_2, S_1) &= L_e(C_1, S_2) + R(C_1) = 2 + 0.1 = 2.1 \\
E_s(C_2, S_1) &= L_e(C_2, S_1) = L_s(C_2, S_1) + L(C_2, S_1) = 2.1 + 1 = 3.1 \\
E_e(C_2, S_1) &= E_s(C_2, S_1) + E(C_2, S_1) = 3.1 + 5 = 8.1
\end{aligned}$$

Since $E_e(C_2, S_1) = 8.1 < E_e(C_1, S_1) = 10$, the constraint (6.3) is not satisfied. In fact, this case will lead to a unideal scheduling.

Therefore, $T_{2\_max} = \min(E_e(C_1, S_2), E_e(C_2, S_1)) - E_s(C_2, S_1) = \min(10, 8.1) - 3.1 = 5$.

- Case 4: $< C_1, S_2 > \rightarrow < C_2, S_2 >$

Let $L_s(C_1, S_2) = 0$, then we have the following equations.

$$\begin{aligned}
L_e(C_1, S_2) &= 2 \\
E_e(C_1, S_2) &= E_s(C_1, S_2) + E(C_1, S_2) = L_e(C_1, S_2) + E(C_1, S_2) \\
&= 2 + 8 = 10 \\
L_s(C_2, S_2) &= L_e(C_1, S_2) + R(C_1) = 2 + 0.1 = 2.1 \\
E_s(C_2, S_2) &= L_e(C_2, S_2) = L_s(C_2, S_2) + L(C_2, S_2) = 2.1 + 2 = 4.1 \\
E_e(C_2, S_2) &= E_s(C_2, S_2) + E(C_2, S_2) = 4.1 + 8 = 12.1
\end{aligned}$$

Since $E_e(C_2, S_2) = 12.1 > E_e(C_1, S_2) = 10$, the constraint (6.3) is satisfied.

Therefore, $T_{2\_max} = E_e(C_1, S_2) - E_s(C_2, S_2) = 10 - 4.1 = 5.9$. Comparing it to the result in the previous case, it is reasonable to speculate that **Speculation (2)** if T_{1_max} is the same, in terms of T_{2_max} the strategy (i.e., S_2) applied to the next concept C_2 with a longer effective duration outperforms the one (i.e., S_1) with shorter effective duration. Under this circumstance, the effective duration of strategies matter again. Hence, we continue to examine more details to solve the inconsistence between **Speculation (1)** and **Speculation (2)**. In addition, from an overall perspective, among the four cases, the T_{2_max} order from the longest to the smallest is: Case 4 (two long-effective strategies) > Case 3 (long-effective strategy prior to short-effective strategy) > Case 1 (two short-effective strategies) > Case 2 (short-effective strategy prior to long-effective strategy). Therefore, how to match strategies and concepts is also a research problem.

6.2.2 An Algebraical Derivation

In this subsection, we use algebraical signs instead of numbers to calculate T_{2_max} of the two-phase learning procedure, i.e., $< C_1, S_i > \rightarrow < C_2, S_j >$.

Again, let $L_s(C_1, S_i) = 0$.

$$\begin{aligned}
L_e(C_1, S_i) &= L_s(C_1, S_i) + L(C_1, S_i) = L(C_1, S_i) \\
E_e(C_1, S_i) &= E_s(C_1, S_i) + E(C_1, S_i) = L_e(C_1, S_i) + E(C_1, S_i) \\
&= L(C_1, S_i) + E(C_1, S_i) \\
L_s(C_2, S_j) &= L_e(C_1, S_i) + R(C_1) = L(C_1, S_i) + R(C_1) \\
E_s(C_2, S_j) &= L_e(C_2, S_j) = L_s(C_2, S_j) + L(C_2, S_j) \\
&= L(C_1, S_i) + L(C_2, S_j) + R(C_1) \\
E_e(C_2, S_j) &= E_s(C_2, S_j) + E(C_2, S_j) \\
&= L(C_1, S_i) + L(C_2, S_j) + E(C_2, S_j) + R(C_1)
\end{aligned}$$

Apparently, $E_s(C_2, S_j) > E_s(C_1, S_i) = L(C_1, S_i)$.
Therefore,

$$T_{2\_\max} = \min(E_e(C_1, S_i), E_e(C_2, S_j)) - E_s(C_2, S_j)$$

When $E_e(C_1, S_i) \leq E_e(C_2, S_j)$ (e.g., case 1, 2, and 4), i.e., **Condition (1)**:

$$E(C_1, S_i) \leq L(C_2, S_j) + E(C_2, S_j) + R(C_1) \tag{6.8}$$

then:

$$\begin{aligned}
T_{2_\max} &= E_e(C_1, S_i) - E_s(C_2, S_j) \\
&= E(C_1, S_i) - L(C_2, S_j) - R(C_1) \leq E(C_2, S_j)
\end{aligned} \tag{6.9}$$

When $E_e(C_1, S_i) > E_e(C_2, S_j)$ (e.g., case 3), i.e., **Condition (2)**:

$$E(C_1, S_i) > L(C_2, S_j) + E(C_2, S_j) + R(C_1) \tag{6.10}$$

then:

$$T_{2\_\max} = E_e(C_2, S_j) - E_s(C_2, S_j) = E(C_2, S_j) \tag{6.11}$$

By the way, if we look backward at Figure 6.1, we can write down the formulas (6.8)–(6.11) sooner. And both the algebraical and line segment derivations show that $T_{2\_\max}$ is irrelevant with $L(C_1, \cdot)$, i.e., the learning duration of C_1.

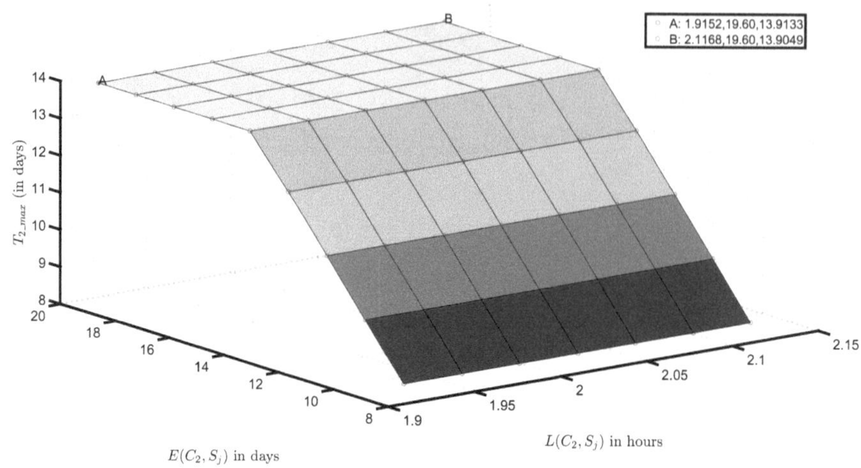

FIGURE 6.2 $T_{2\_\max}$ surface when $R(C_1) = 10$ minutes and $E(C_1, S_i) = 14$ days.

6.2.3 $T_{2\_\max} = f(L(C_2, S_j), E(C_2, S_j))$ Surface

According to the different expressions of $T_{2\_\max}$ under different conditions, i.e., Equations (6.8)–(6.11),

$$T_{2\_\max} = \begin{cases} E(C_2, S_j), E(C_1, S_i) > L(C_2, S_j) + E(C_2, S_j) + R(C_1) \\ E(C_1, S_i) - L(C_2, S_j) - R(C_1), \text{otherwise} \end{cases} \tag{6.12}$$

By Equation (6.12), $T_{2\_\max}$ is essentially affected by four variables: $L(C_2, S_j)$, $E(C_2, S_j)$, $E(C_1, S_i)$, and $R(C_1)$. It is impossible to reflect all the five dimensions in a 3D cubic. However, we can control two less important variables, i.e., $E(C_1, S_i)$ and $R(C_1)$. The following figures present the surfaces of $T_{2\_\max} = f(L(C_2, S_j), E(C_2, S_j))$ when $R(C_1)$ is controlled at 10 minutes, 1 hour, 0.5 day, and 1 day, and $E(C_1, S_i)$ is fixed at 14 days (2 weeks). The last value comes from the length of an observation window in Yang and Chen (2021), which does not necessarily mean that there indeed exists a 14-days-effective strategy. We adopt this value because it is probably at the same scale as a real one.

As shown in Figures 6.2–6.5, the $T_{2\_\max}$ surface exhibits a pattern: "ramp-platform". The turning "line" or the "ridge" appears when $E(C_2, S_j) = 14$ days. On a macro trend, for a given $L(C_2, S_j)$, $T_{2\_\max}$ increases as $E(C_2, S_j)$ increases at the beginning (approximately from 9 days to 14 days), because it follows the rule of Equation (6.11). After it

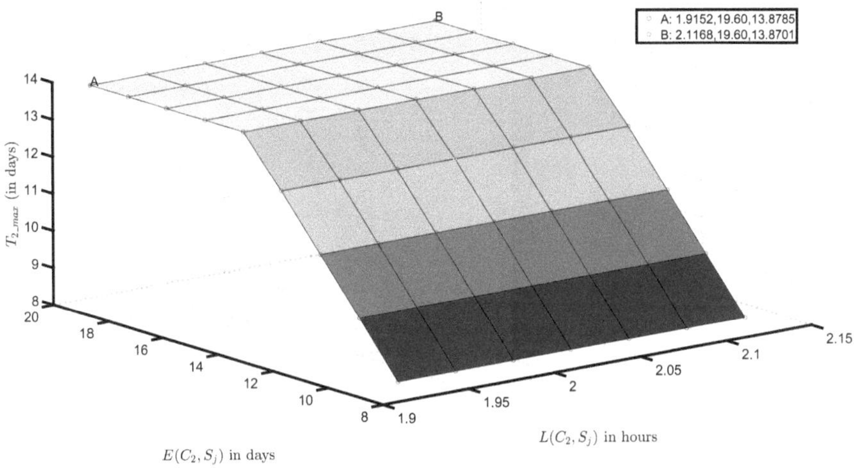

FIGURE 6.3 $T_{2\_\max}$ surface when $R(C_1) = 60$ minutes and $E(C_1, S_i) = 14$ days.

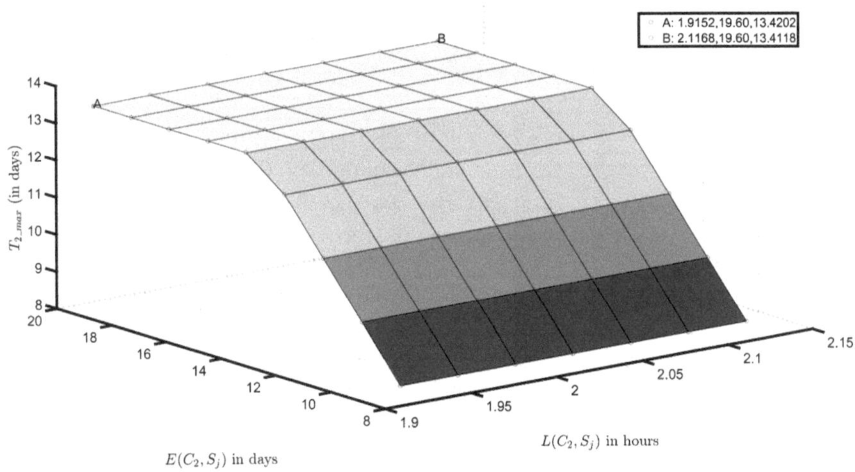

FIGURE 6.4 $T_{2\_\max}$ surface when $R(C_1) = 0.5$ day and $E(C_1, S_i) = 14$ days.

"climbs" over the "ridge", it turns into a platform because $E_e(C_1, S_i) \leq E_e(C_2, S_j)$ and it turns to follow the rule of Equation (6.9). On a micro trend, however, the platform is not strictly horizontal. In each one of Figures 6.2–6.5, we mark out two points on the surface: A and B, and list their coordinates in the legend area. The coordinates are in the format: $L(C_2, S_j)$, $E(C_2, S_j)$, and $T_{2\_\max}$, from the left to the right.

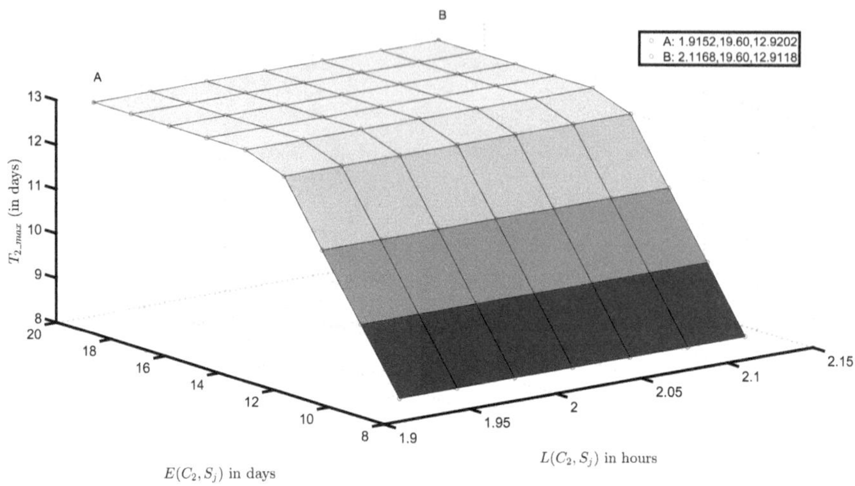

FIGURE 6.5 $T_{2\_\max}$ surface when $R(C_1) = 1$ day and $E(C_1, S_i) = 14$ days.

As seen, for a given $E(C_2, S_j)$ in the platform region, $T_{2\_\max}$ decreases as $L(C_2, S_j)$ increases, though such change is very subtle. Therefore, by plotting the $T_{2\_\max}$ surface, the seemingly inconsistence between **Speculation (1)** and **Speculation (2)** is solved by the surface's geometry characteristics.

In addition, the platform's height $(\max(T_{2\_\max}))$ decreases as $R(C_1)$ increases, which is consistent with our intuition. And the decreasing amplitude is also subtle unless $R(C_1)$ is longer than 1 day, indicating that an appropriate rest will not shorten the overall effective duration significantly.

6.3 GENERAL SCHEDULING

In the preface of this chapter, the ideal scheduling is defined as a scheduling where each invalid moment (ending moment of the effective duration) of a concept is later than the one of its precedent concepts. However, this is not the common case. The common case is that a learner may start to forget some precedent concept when the one starts to learn its successors or the successors of its successors. In this section, we try to investigate a more general case that includes the above phenomenon, i.e., general scheduling.

TABLE 6.2 Notations

Notations	Description
C	Concept index. In a mathematical course, the concepts are presented in an order, which cannot be altered casually. For example, learners have to learn addition before multiplication. Therefore, the index of a mathematical concept is its presenting order in the course. For example, $C = 3$ indicates the third concept in a particular concept presenting/instructional sequence of the course.
$S(C)$	Strategy set for the C-th concept. Normally, there may exist more than one strategy to learn and master a concept.
$L_s(c, s)$	Starting moment of learning concept c by strategy s.
$L_e(c, s)$	Ending moment of learning concept c by strategy s.
$E_s(c, s)$	Starting moment of the effective duration when learning concept c by strategy s.
$E_e(c, s)$	Ending moment of the effective duration when learning concept c by strategy s.
$L(c, s)$	Learning duration for concept c using strategy s.
$E(c, s)$	Effective duration for concept c using strategy s.
$R(c)$	Rest duration between the moment after learning concept c and the moment to learn the next concept.
$< c, s >$	A phase in a learning procedure: learning concept c with strategy s.
T_{i_max}	The maximum intersecting effective duration for concept 1 to i, after the i-th concept is mastered.

6.3.1 Exhaustive Search

Before we start, we list the notations in Table 6.2. Most of them have been used by the previous sections. For simplicity, we use a concept index (more details can be found in Table 6.2), i.e, C, instead of C_i, to denote a concept. This is the only difference from the previous sections.

We design an exhaustive search algorithm to find $T_{\max}$ where there are only two and three concepts, respectively. The basic idea is that the time line of each action in a learning procedure (i.e., start to learn, end of learning, rest, start to be effective, and end of being effective) is accumulated concept by concept. Line 21 tells the connections between current concept and its precedent concept. Lines 22–24 establish the relationships between various parameters about current concept by their definitions. In fact, Line 11 and Lines 12–14 function in the same way. However, we lack of a convenient and easy to understand way to integrate them in one place. Readers may follow the patterns exhibited by these pseudo codes to extend this algorithm into a four-concept or multiple-concept scenario easily.

Algorithm 6.1: Exhaustive Search

Input: Learning duration matrix: $L(concept, strategy)$;
Effective duration matrix: $E(concept, strategy)$;
Rest duration vector: $R(concept)$

Output: Intersecting effective duration $T_{\max}$ for each learning procedure π (e.g., <concept $(x-1)$, strategy l> $\rightarrow$ <concept x, strategy m> $\rightarrow$ <concept $(x+1)$, strategy n>)

Initialize **L**, **E** and **R**;
$L_s(1, 1:N) \leftarrow 0$;
$C \leftarrow 1$;
foreach *Strategy i ∈ S(C)* **do**
  $L_e(C,i) \leftarrow L_s(C,i) + L(C,i)$;
  $E_s(C,i) \leftarrow L_e(C,i)$;
  $E_e(C,i) \leftarrow L_e(C,i) + E(C,i)$;
end
$C \leftarrow 2$;
foreach *Strategy j ∈ S(C)* **do**
  foreach *Strategy i ∈ S(C-1)* **do**
    $L_s(C,j) \leftarrow L_e(C-1,i) + R(C-1)$;
    $L_e(C,j) \leftarrow L_s(C,j) + L(C,j)$;
    $E_s(C,j) \leftarrow L_e(C,j)$;
    $E_e(C,j) \leftarrow E_s(C,j) + E(C,j)$;
    $T_{2\_\max} = \min(E_e(C-1,i), E_e(C,j)) - \max(E_s(C-1,i), E_s(C,j))$
    print $< C-1, i > \rightarrow < C, j >$: $T_{2\_\max}$
  end
end
$C \leftarrow 3$;
foreach *Strategy k ∈ S(C)* **do**
  foreach *Strategy j ∈ S(C-1)* **do**
    foreach *Strategy i ∈ S(C-2)* **do**
      $L_s(C,k) \leftarrow L_e(C-1,j) + R(C-1)$;
      $L_e(C,k) \leftarrow L_s(C,k) + L(C,k)$;
      $E_s(C,k) \leftarrow L_e(C,k)$;
      $E_e(C,k) \leftarrow E_s(C,k) + E(C,k)$;
      $T_{3\_max} = \min(E_e(C-2,i), E_e(C-1,j), E_e(C,k)) - \max(E_s(C-2,i), E_s(C-1,j), E_s(C,k))$
      print $< C-2, i > \rightarrow < C-1, j > \rightarrow < C, k >$: T_{3_max}
    end
  end
end

If the parameters in Table 6.1 are adopted by the algorithm as inputs, it will output the same results calculated manually in cases one to four in the two-concept scenario. In a three-concept scenario, when the inputs are the same as the two-concept scenario, the output is listed in Table 6.3.

The table can also be presented in a tree form in Figure 6.6. And the tree view can deliver visualized messages as follows.

TABLE 6.3 Outputs of Three-Concept Learning

Phase 1	**Phase 2**	**Phase 3**	T_{max}
$< 1, 1 >$			5.0
$< 1, 1 >$	$\to < 2, 1 >$		3.9
$< 1, 1 >$	$\to < 2, 1 >$	$\to < 3, 1 >$	1.8
$< 1, 1 >$	$\to < 2, 1 >$	$\to < 3, 2 >$	0.8
$< 1, 1 >$	$\to < 2, 2 >$		2.9
$< 1, 1 >$	$\to < 2, 2 >$	$\to < 3, 1 >$	0.8
$< 1, 1 >$	$\to < 2, 2 >$	$\to < 3, 2 >$	−0.2
$< 1, 2 >$			8.0
$< 1, 2 >$	$\to < 2, 1 >$		5.0
$< 1, 2 >$	$\to < 2, 1 >$	$\to < 3, 1 >$	3.9
$< 1, 2 >$	$\to < 2, 1 >$	$\to < 3, 2 >$	2.9
$< 1, 2 >$	$\to < 2, 2 >$		5.9
$< 1, 2 >$	$\to < 2, 2 >$	$\to < 3, 1 >$	4.8
$< 1, 2 >$	$\to < 2, 2 >$	$\to < 3, 2 >$	3.8

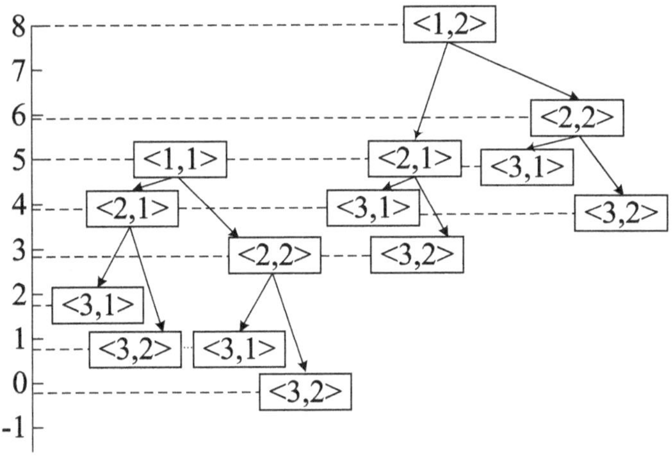

FIGURE 6.6 Tree view of Table 6.3.

(1) The intersecting effective duration (e.g. T_{2_max}, T_{3_max}) decreases as the number of phases (or number of concepts learned) increases. One leaf ($< 3, 2 >$ in the learning procedure $< 1, 1 > \to < 2, 2 > \to < 3, 2 >$) in the tree view even appears negative (−0.2). In fact, the purpose of leaving such non-sense result both in Table 6.3 and Figure 6.6 is to remind readers that an inappropriate scheduling may exhaust the intersecting effective duration slowly reduced previously so fast that no precedent concept is retained before their following concept is about to be learned. Such disaster may

be tolerable in ill-structured courses, but fatal for well-structured courses, e.g., maths. For example, it may correspond to a situation that a learner may forget addition before the one is about to learn multiplication. If this phenomenon happens, what the learner is going to do next is not learning multiplication but reviewing addition.

(2) The eye-catchy learning procedure in the tree view is $< 1, 2 > \rightarrow < 2, 2 > \rightarrow < 3, 1 >$, because it has the longest intersecting effective duration not only in the end but also for every kinds of learning procedures (i.e., one-concept learning, two-concept learning, and three-concept learning). It also reflects our aforementioned recommendation: in most cases, if there are multiple strategies for a concept, select the strategy whose effective duration is the longest, and such selection may benefit the final intersecting effective duration. However, the last phase $< 3, 1 >$ violates this recommendation. A detailed examination reveals that this is because though strategy 1 has a shorter effective duration than strategy 2, the former also has a shorter learning duration than the latter. Shorter learning duration can make the strategy take effect sooner than the longer one. This (i.e., taking effect sooner) is particularly helpful, when the effective duration of a newly learned concept is becoming less helpful to mitigate the trend that current intersecting effective duration is becoming shorter. By this means, phase $< 3, 1 >$ "contributes more" (reduces less, actually) intersecting effective duration than phase $< 3, 2 >$ after the learning procedure: $< 1, 2 > \rightarrow < 2, 2 >$.

Suppose $|S(1)| = I, |S(2)| = J, |S(3)| = K$, i.e., the numbers of strategies for learning the first, second, and third concepts, are I, J, and K, respectively. The time complexity of Algorithm 6.1 is $\mathcal{O}(IJK)$. Based on this three-concept case, we can easily extrapolate that the time complexity of the exhaustive search algorithm will be approximately $\mathcal{O}(H^N)$ if there are N concepts ($N > 3$) and $H = \max(|S(1)|, |S(2)|, \ldots, |S(N)|)$. In the next subsection, an iterative search algorithm will be proposed to reduce this exponential time complexity to polynomial time complexity.

6.3.2 Iterative Search

One drawback of Algorithm 6.1 is that it does not examine whether there is a recurrence relation between the current $T_{\max}$ and the previous one in each loop, such as Equations (6.6) and (6.7) in Section 6.1. If there is, many combinations of concepts and strategies or learning sequences, which do not contribute to the optimal learning procedure maximizing the intersecting effective duration, will be abandoned early. This will save a lot of time, compared to the original algorithm, though there are still multiple recurrence relations for general scheduling scenarios to discuss, which is unlike the single recurrence relation in the ideal scheduling scenario.

In this subsection, we will first examine if there exists a recurrence relation between the current $T_{\max}$ and the previous one in each loop, or the recurrence relation between their candidates among which they are possibly from. The following theorem answers this question.

Theorem 6.1. *In an n-concept learning, a learner learns one concept by one concept. In a mathematical course, the learning orders about the concepts are fixed, but the strategies are not. Applying different learning strategies (s) to a concept (c) may lead to different effective durations. In addition, all the rest duration between adjacent two learning phases are assumed to be constant for any strategy. If the global intersecting effective duration (T) covering all the n concepts can be positive, i.e., the learner can learn all the concepts once and retain all of them for some time, the learning procedure which maximizes T can be constructed by the phases $< c, s >$ which make the local intersecting effective durations also the maximum.*

Note that local intersecting effective duration of the learning sequence $< 1, i > \to < 2, j >$ is denoted by $T(< 1, i > \to < 2, j >)$, and,

$$T(< 1, i > \to < 2, j >) = min(E_e(1, i), E_e(2, j)) - max(E_s(1, i), E_s(2, j)).$$

Local intersecting effective duration of the learning sequence $< 1, i > \to < 2, j > \to < 3, k >$ is denoted by $T(< 1, i > \to < 2, j > \to < 3, k >)$ and

$$T(< 1, i > \to < 2, j > \to < 3, k >) = \begin{aligned}[t] &min(E_e(1, i), E_e(2, j), E_e(3, k)) - \\ &max(E_s(1, i), E_s(2, j), E_s(3, k)). \end{aligned}$$

and so on.

Proof. *This proof adopts Mathematical Induction.*

Suppose $T(1, l) = \max_{p \in S(1)} E(1, p)$. If the mathematical course has only one concept, the theorem is proved.

If the mathematical course has more than one concept, no matter what strategy is applied to the second concept, $T(< 1, p > \to < 2, j >)$ will not be the largest unless p is replaced by l.

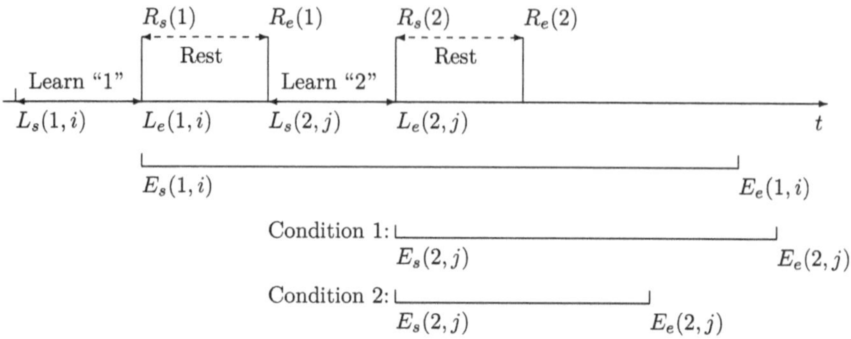

FIGURE 6.7 Reviewing the two conditions (for brevity "1" is for the first concept and "2" is for the second concept.)

This is because $T(<1,p>\to<2,j>) = E(1,p) - \Delta(2,j)$. $\Delta(2,j)$ *denotes the loss of* $E(1,p)$ *caused by* $<(2,j>$ *when it follows* $<1,p>$ *into the learning procedure.*

$\Delta(2,j) = R(1) + L(2,j)$ *when* $E_e(2,j) \geq E_e(1,p)$ *(see* ***Condition (1)****, i.e., Equation (6.8) for (6.9)), or*

$\Delta(2,j) = E(1,p) - E(2,j)$ *when* $E_e(2,j) < E_e(1,p)$ *(see* ***Condition (2)****, i.e., Equation (6.10) for (6.11)).*

For the former case, $\Delta(2,j)$ *is irrelevant to either* $E(1,\cdot)$ *or* $L(1,\cdot)$*, and* $R(1)$ *is assumed to be constant for any strategy in our current theorem; therefore,* $T(<1,p>\to<2,j>)$ *will be maximum for every* $<2,j>$*, if and only if* $E(1,p)$ *is the minimum.*

For the latter case, $\Delta(2,j)$ *is relevant to* $E(1,p)$*; however,*

$T(<1,p>\to<2,j>) = E(1,p) - (E(1,p) - E(2,j)) = E(2,j)$ *now, which is also irrelevant to either* $E(1,\cdot)$ *or* $L(1,\cdot)$*,* $E(1,p)$*'s being maximum under the condition* $E_e(2,j) < E_e(1,p)$ *will not break this equation.*

In sum, $\max\limits_{p\in S(1)} T(<1,p>\to<2,j>) = T(1,l) - \Delta(2,j)$*. And* $\arg\max_{p\in S(1)} T(<1,p>\to<2,j>) =<1,l>\to<2,j>$.

From now on, we are going to investigate the recurrence relation between the local intersecting effective duration of the learning sequence $<1,p>\to \dots \to<m-1,k>$*, denoted by* $T(m-1)$ *(*$m \geq 2$*), and the one of the learning sequence* $<1,p>\to \cdots \to<m,j>$*), denoted by* $T(m)$*. Suppose* $T(m)$ *starts at the moment* $T_s(m)$ *and ends at the moment* $T_e(m)$*,*

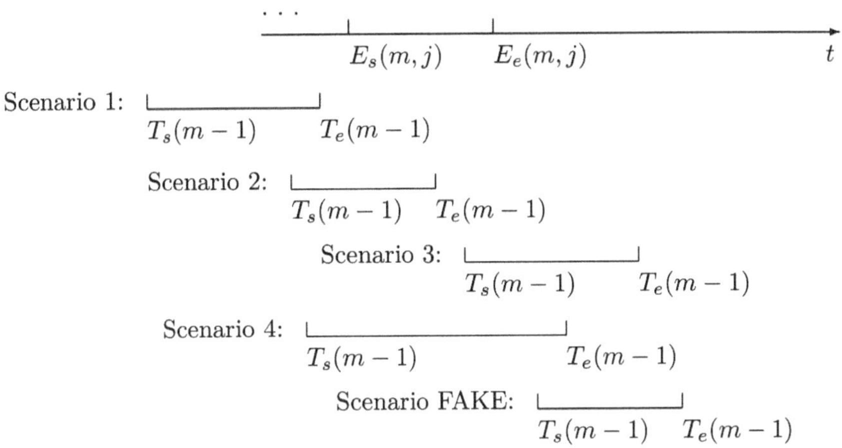

FIGURE 6.8 Four scenarios.

i.e., $T(m) = T_e(m) - T_s(m)$. *And,*

$$[T_s(m), T_e(m)] = [T_s(m-1), T_e(m-1)] \cap [E_s(m,j), E_e(m,j)] \quad (6.13)$$

The relative positional relationship between the time interval $[T_s(m-1), T_e(m-1)]$ *and the time interval* $[E_s(m), E_e(m)]$ *has four kinds of scenarios. They are presented in Figure 6.8, which also includes a fake kind of scenario. The fake kind of scenario will be proven not to occur in the next proposition. In the rest of this proof, we will discuss the recurrence relation between* $T(m-1)$ *and* $T(m)$ *with the help of Equation (6.13) in the four kinds of scenarios, namely, Scenarios 1, 2, 3 and 4, respectively.*

Scenario 1: This scenario occurs when $T_e(m-1) < E_s(m,j)$. *In this kind of scenario,* $[T_s(m), T_e(m)] = [T_s(m-1), T_e(m-1)] \cap [E_s(m,j), E_e(m,j)] = \emptyset$. *Therefore,* $T(m) = 0$. *And a larger* $T(m-1)$ *will not affect* $T(m) = 0$ *as long as* $T(m-1)$ *is constrained by* $T_e(m-1) < E_s(m,j)$.

Scenario 2: This scenario occurs when $T_s(m-1) \leq E_s(m,j)$ *and* $E_s(m,j) \leq T_e(m-1)$ *and* $T_e(m-1) \leq E_e(m,j)$. *In this kind of scenario,* $[T_s(m), T_e(m)] = [T_s(m-1), T_e(m-1)] \cap [E_s(m,j), E_e(m,j)] = [Es(m,j), T_e(m-1)]$. *Therefore,* $T(m) = T_e(m-1) - E_s(m,j)$. *By this equation, a larger* $T(m)$ *requires a larger* $T_e(m-1)$. *And a larger* $T_e(m-1)$ *leads to a larger* $T(m-1)$. *Therefore, a larger* $T(m)$ *requires a larger* $T(m-1)$.

Scenario 3: This scenario occurs when $E_s(m,j) \leq T_s(m-1)$ *and* $T_s(m-1) \leq E_e(m,j)$ *and* $E_e(m,j) \leq T_e(m-1)$. *In this kind of scenario,* $[T_s(m), T_e(m)] = [T_s(m-1), T_e(m-1)] \cap [E_s(m,j), E_e(m,j)] = [T_s(m-1), E_e(m,j)]$. *Therefore,* $T(m) = E_e(m,j) - T_s(m-1)$. *By this equation, a larger* $T(m)$ *requires a smaller* $T_s(m-1)$. *And a smaller* $T_s(m-1)$ *leads to a larger* $T(m-1)$. *Therefore, a larger* $T(m)$ *requires a larger* $T(m-1)$.

Scenario 4: This scenario occurs when $T_s(m-1) < E_s(m,j)$ *and* $T_e(m-1) > E_e(m,j)$. *In this kind of scenario,* $[T_s(m), T_e(m)] = [T_s(m-1), T_e(m-1)] \cap [E_s(m,j), E_e(m,j)] = [E_s(m,j), E_e(m,j)]$. *Therefore,* $T(m) = E(m,j)$. *Note that a larger* $T(m-1)$ *will not affect this equation.*

Therefore,$max_{j \in S(m)} T(m)$ *will be found in a learning procedure* $< 1, p > \to \ldots \to < m, j >$) *which contains a sub-procedure* $< 1, p > \to \ldots \to < m-1, k >$) *maximizing* $T(m-1)$. *And this is just the conclusion of this theorem.*

□

Proposition 1. *Scenario FAKE in Figure 6.8 will not occur in the physical world.*

Proof. *It can be proved by contradiction. Suppose Scenario FAKE in Figure 6.8 can happen in the physical world. Then, according to the positional relationship between* $T_s(m-1)$ *and* $E_e(m,j)$, *we have* $T_s(m-1) > E_e(m,j)$. *Considering* $T_s(m-1) = \max(E_s(m-1,i), E_s(m-2,j), \ldots, E_s(1,z))$, *we assume* $\max(E_s(m-1,i), E_s(m-2,j), \ldots, E_s(1,z)) = E_s(x,y)$ $(1 \leq x \leq m-1)$. *Then we have* $T_s(m-1) = E_s(x,y) > E_e(m,j)$. *Note that* $E_s(x,y) = L_e(x,y)$, *we have* $L_e(x,y) > E_e(m,j)$. *Since* $E_e(m,j) > E_s(m,j) = L_e(m,j)$, *we finally have* $L_e(x,y) > L_e(m,j)$. *On the contrary, note that* $x \leq m-1$, *which means concept* x *should be learnt before* m. *Therefore, the ending moment of learning concept* x, *i.e.,* $L_e(x,y)$ *should be earlier than the ending moment of learning concept* m, *i.e.,* $L_e(m,j)$, *that is,* $L_e(x,y) < L_e(m,j)$. *This is a contradiction to the aforementioned* $L_e(x,y) > L_e(m,j)$.

The pseudo codes of the following iterative search algorithm (i.e., Algorithm 6.2) inherit the notations listed in Table 6.2, and also include several additional notations as in Table 6.4.

The basic idea of iterative search is that the recurrence relation between $T_{C\_\max}$ and $T_{(C-1)\_\max}$ is established by examining the relationships

TABLE 6.4 Supplemented Notations to Table 6.2 for Algorithm 6.2

Notations	Description
$T(C, s)$	One of candidates at round C (also the C-th concept), among which $T_{C\_\max}$ is found.
$T_s(C, s)$	Starting moment of $T(C, s)$.
$T_e(C, s)$	Ending moment of $T(C, s)$.
$T_{i\_\max\_s}$	Starting moment of $T_{i\_\max}$ in Table 6.2.
$T_{i\_\max\_e}$	Ending moment of $T_{i\_\max}$ in Table 6.2.

Algorithm 6.2: Iterative Search

Input: Learning duration matrix: $L(concept, strategy)$;
Effective duration matrix: $E(concept, strategy)$;
Rest duration vector: $R(concept)$

Output: Intersecting effective duration $T_{\max}$ for each learning procedure π

Initialize **L**, **E** and **R**;
foreach *Strategy* $s \in S(1)$ **do**
    $L_s(1, s) \leftarrow 0; L_e(1, s) \leftarrow L_s(1, s) + L(1, s); E_s(1, s) \leftarrow L_e(1, s);$
    $E_e(1, s) \leftarrow E_s(1, s) + E(1, s);\ T_s(1, s) \leftarrow E_s(1, s);$
    $T_e(1, s) \leftarrow T_s(1, s) + E(1, s);\ T(1, s) \leftarrow T_e(1, s) - T_s(1, s);$
end
$T_{1\_\max} \leftarrow \max_{s \in S(1)} T(1, s);\ s_{\text{opt}} \leftarrow \arg\max_{s \in S(1)} T(1, s);$
$T_{\max\_s}(1) \leftarrow T_s(1, s_{\text{opt}});\ T_{\max\_e}(1) \leftarrow T_e(1, s_{\text{opt}});$
$\text{Seq}_{\text{opt}}(1) \leftarrow [T_{\max\_s}(1), T_{\max\_e}(1), T_{1\_\max}, 1, s_{\text{opt}}];$
for $C \leftarrow 2$ **to** N **do**
    foreach *Strategy* $s \in S(C)$ **do**
        $L_e(C, s) \leftarrow L_e(C-1, s_{\text{opt}}) + R(C-1) + L(C, s);$
        $E_s(C, s) \leftarrow L_e(C, s);\ E_e(C, s) \leftarrow E_s(C, s) + E(C, s);$
        if $T_{\max\_e}(C-1) < E_s(C, s)$ **then**
            $T(C, s) \leftarrow 0;$
        else if $T_{\max\_s}(C-1) \leq E_s(C, s) \leq T_{\max\_e}(C-1) \leq E_e(C, s)$ **then**
            $T_s(C, s) \leftarrow E_s(C, s);\ T_e(C, s) \leftarrow T_{(C-1)\_\max\_e};$
            $T(C, s) \leftarrow T_e(C, s) - T_s(C, s);$
        else if $E_s(C, s) \leq T_{(C-1)\_\max\_s} \leq E_e(C, s) \leq T_{(C-1)\_\max\_e}$ **then**
            $T_s(C, s) \leftarrow T_{(C-1)\_\max\_s};\ T_e(C, s) \leftarrow E_e(C, s);$
            $T(C, s) \leftarrow T_e(C, s) - T_s(C, s);$
        else if $T_{(C-1)\_\max\_s} \leq E_s(C, s)$ **and** $E_e(C, s) \leq T_{(C-1)\_\max\_e}$ **then**
            $T_s(C, s) \leftarrow E_s(C, s);\ T_e(C, s) \leftarrow E_e(C, s);$
            $T(C, s) \leftarrow E(C, s);$
    end
    $T_{C\_\max} \leftarrow \max_{s \in S(C)} T(C, s);\ s_{\text{opt}} \leftarrow \arg\max_{s \in S(C)} T(C, s);$
    $T_{\max\_s}(C) \leftarrow T_s(C, s_{\text{opt}});\ T_{\max\_e}(C) \leftarrow T_e(C, s_{\text{opt}});$
    $\text{Seq}_{\text{opt}}(C) \leftarrow [T_{\max\_s}(C), T_{\max\_e}(C), T_{C\_\max}, C, s_{\text{opt}}];$
end

TABLE 6.5 Output of Iterative Search for Four-Concept Learning

Concept Index	Strategy Index	T_{max_s}	T_{max_e}	T_{max}(Concept Index)
1	2	2.0	10.0	8.0
2	2	4.1	10.0	5.9
3	1	5.2	10.0	4.8
4	1	6.3	10.0	3.7

between their candidates, i.e., $T(C, s)$ and $T(C - 1, s')$. Specially, such relationships are discussed in four scenarios, i.e., Lines 14 and are for Scenario 1, Lines 16–18 are for Scenario 2, Lines 19–21 are for Scenario 3, and Lines 22–24 are for Scenario 4.

The input for this algorithm is still the parameters in Table 6.1, but extended into a four-concept learning situation. Its output is presented in Table 6.5. Therefore, the optimal learning procedure is $< 1, 2 > \rightarrow < 2, 2 > \rightarrow < 3, 1 > \rightarrow < 4, 1 >$. And the maximum intersecting effective duration for concepts 1 to 4 is 3.7.

The output of exhaustive search (i.e., Algorithm 6.1) with the same input is presented in Table 6.6, which still keeps the negative results. The row that contains the maximum $T_{\max}$ is boxed out. This row is consistent with the result presented in Table 6.5. So, the outputs of exhaustive search and iterative search both deliver the same optimal solution; however, the latter one is certainly much faster. If there are N concepts ($N > 3$) and $H = \max(|S(1)|, |S(2)|, \ldots, |S(N)|)$, the time complexity of the latter algorithm is decreased to approximately $\mathcal{O}(NH)$ from approximately $\mathcal{O}(H^N)$.

6.4 SCHEDULING FOR REVIEWING

Sections 6.1–6.3 focus on how to schedule the learning strategies so that the concepts can be retained as long as possible, given that learning each concept must be separated by a blank time for rest. By using Algorithm 6.1 or 6.2, we have the maximum intersecting effective duration, corresponding to an optimal learning procedure ($< 1, 2 > \rightarrow < 2, 2 > \rightarrow < 3, 1 >$), when there are three concepts and their learning/effective durations with different strategies are listed in Table 6.1. Figure 6.9 illustrates a scheduling chart, whose left part is a detailed version of the optimal learning procedure in Table 6.3 or Figure 6.6. Its right part is for a second time learning, i.e., reviewing.

Some assumptions and context of scheduling for reviewing are introduced in this section first, and then how to read the scheduling chart

TABLE 6.6 Output of Exhaustive Search for Four-Concept Learning

Phase 1	**Phase 2**	**Phase 3**	**Phase 4**	T_{max}
$< 1,1 >$	$< 2,1 >$	$< 3,1 >$	$< 4,1 >$	−0.3
$< 1,2 >$	$< 2,1 >$	$< 3,1 >$	$< 4,1 >$	1.8
$< 1,1 >$	$< 2,2 >$	$< 3,1 >$	$< 4,1 >$	−0.3
$< 1,2 >$	$< 2,2 >$	$< 3,1 >$	$< 4,1 >$	3.7
$< 1,1 >$	$< 2,1 >$	$< 3,2 >$	$< 4,1 >$	−1.3
$< 1,2 >$	$< 2,1 >$	$< 3,2 >$	$< 4,1 >$	0.8
$< 1,1 >$	$< 2,2 >$	$< 3,2 >$	$< 4,1 >$	−1.3
$< 1,2 >$	$< 2,2 >$	$< 3,2 >$	$< 4,1 >$	2.7
$< 1,1 >$	$< 2,1 >$	$< 3,1 >$	$< 4,2 >$	−1.3
$< 1,2 >$	$< 2,1 >$	$< 3,1 >$	$< 4,2 >$	0.8
$< 1,1 >$	$< 2,2 >$	$< 3,1 >$	$< 4,2 >$	−1.3
$< 1,2 >$	$< 2,2 >$	$< 3,1 >$	$< 4,2 >$	2.7
$< 1,1 >$	$< 2,1 >$	$< 3,2 >$	$< 4,2 >$	−2.3
$< 1,2 >$	$< 2,1 >$	$< 3,2 >$	$< 4,2 >$	−0.2
$< 1,1 >$	$< 2,2 >$	$< 3,2 >$	$< 4,2 >$	−2.3
$< 1,2 >$	$< 2,2 >$	$< 3,2 >$	$< 4,2 >$	1.7

(Figures 6.9 and 6.10) is illustrated. The implications from these two numeric examples (presented in these two scheduling charts) are discussed at last.

6.4.1 Assumptions and Context

The assumptions are claimed in the preface of this chapter. They are rephrased in an item list as follows.

1. Limited effective duration assumption—That is, each learning strategy has a limited effective duration. After this duration, the concept learned will be completely forgotten as if it has never learned.
2. Jenga assumption—That is, when a concept is completely forgotten, all of its descendant concepts are either forgotten as well or not allowed to be learned. For example, if a learner has forgotten addition, what he or she is going to do next is not to learn multiplication, but to review addition. This assumption is proposed specially for well-structured courses, such as maths. Jenga is a game where players take turns to pull a piece of wood from a tower constructed by wood, to see whose pulling makes the tower collapse. Here we compare Jenga to a well-structured course, and compare the pieces of wood to concepts in this course, and compare pulling out a piece of wood to forgetting a concept, and compare the tower's collapsing to the

forgotten concept's all descendant concepts' invalidation. The only difference is that pulling a piece of wood from the tower may not affect a collapsing immediately, whereas forgetting a concept will affect all of its descendant concepts immediately.
3. Spacing assumption—That is, between adjacent intervals of learning, there is a spacing time for rest. However, this constraint will be relaxed for reviewing when necessary.

The context for review is that the optimal scheduling of learning strategies for the first time learning is solved in Sections 6.1–6.3. The maximum intersecting effective duration is 4.8 as shown in Table 6.3 or Figure 6.6. However, such duration is still limited. The question now turns into how to schedule strategies for reviewing so that the intersecting effective duration can be extended as long as possible? Will the scheduling for the second time learning is the same with the one for the first time learning?

We are not going too far or too deep in this direction (otherwise a new book is required), but to report a pilot study to interest any latecomers. Therefore, we stipulate that there is only one review strategy to narrow down the scope of this pilot study. And, to reveal the essentials of the problem proposed in the last paragraph fast, we propose a set of numeric conditions to make the pilot study simpler, rather than formalize it and deliver a comprehensive solution. The set of numeric conditions is that when a learner starts to review a concept, the review duration of the second time learning is half of the learning duration of the first one, and the effective duration of the second time learning is twice of the effective duration of the first one.

Specifically, in the optimal scheduling for the first time learning, two of three concepts are learned with a strategy whose learning duration is 2 and the corresponding effective duration is 8. We stipulate that this strategy will be also used in the second time learning, i.e., reviewing. In addition, due to its second use, the reviewing duration is half of 2, i.e., 1 time unit, and the effective duration is double of 8, i.e., 16 time units.

6.4.2 How to Read the Charts

In this subsection, how to read the scheduling charts is illustrated. The first figure, i.e., Figure 6.9 is also used as an example for the legends. And then,

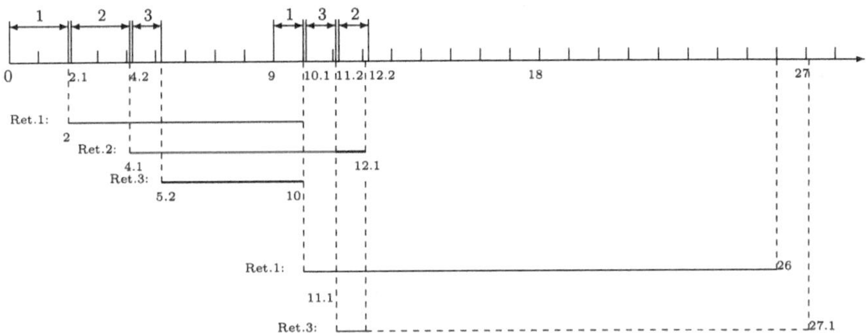

FIGURE 6.9 Scheduling for reviewing when resting is required.

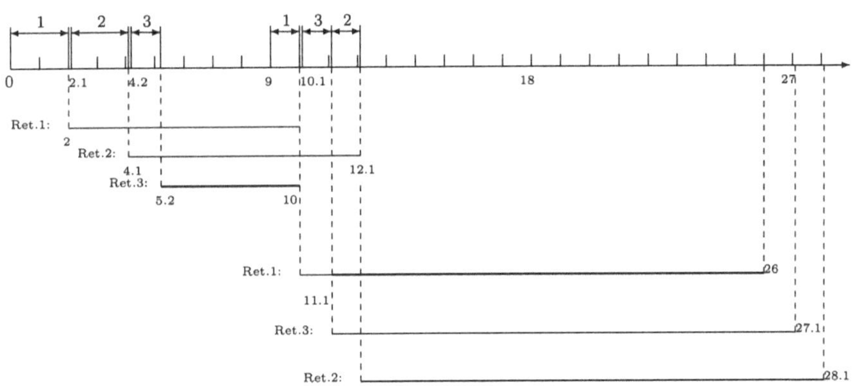

FIGURE 6.10 Scheduling for reviewing when resting is not required.

two stories visualized by these two charts in this subsection are described in text.

In Figure 6.9, the longest horizontal axis with a right arrow is a time axis. The time ticks are distributed evenly from 0 to 28 with a certain time unit. On this time axis, there are two groups of three cabins with two small slots. A cabin with a double arrow on the top indicates the time interval for a concept learning. Therefore, the first group of cabins is for the first time learning, and the second group is for the reviewing. The numbers above the "rooftop" (double arrow) of a cabin indicate the concept indices. For example, the time interval for learning concept 1 at the first time is $[0, 2]$, and the time interval for reviewing concept 1 is $[9, 10]$. Therefore, the concept sequence in the learning group is learning concept 1 first, concept 2 second, and concept 3 finally. Why the concept sequence in the reviewing group is concept 1, concept 3 and then concept 2 will be explained later. When two adjacent cabins have a small slot in between, there is blank time for a rest, lasting for 0.1 time unit, which follows the parameter in Table 6.1.

Beneath the time axis, there are effective intervals for each concept learning or reviewing. Ahead of these effective intervals are their labels. For example, "Ret. 1" is the effective interval for concept 1. "Ret." is short for "retention", meaning it is a retention interval for concept 1. We use "Ret." for "retention" instead of "Eff." for "effectiveness", simply because the former may probably make readers recall its full meaning fast. The solid horizontal line is the real interval for a certain concept, while the dashed horizontal line is the expected interval for a certain concept, and it is not real due to some constraints. The thick solid horizontal line is the maximum intersecting effective interval for a group of three-concept learning. The vertical dashed line is for the moment alignment to the time axis. The left vertical dashed line and the right vertical dashed line attached to a horizontal solid line are like a pair of ropes for a swing, indicating the beginning and ending moments of an effective interval, respectively. For example, in the first learning, the effective (retention) interval for concept 1, 2, and 3 is $[2, 10]$, $[4.1, 12.1]$[1] and $[5.2, 10]$,[2] respectively.

Now, the detailed illustration on the scheduling chart for the learning intervals and effective intervals of the first group is finished. The first group is also the optimal learning procedure found by Algorithm 6.1 or 6.2. Next, we are going to adopt a lazy and adaptive paradigm to schedule the reviewing, aiming to extend the maximum intersecting effective intervals.

The "lazy" paradigm means that we do not make a move until it is necessary. In Figure 6.9, "Ret. 1" at the first learning ends at 10. If the learner does not re-learn (review) it at 9 to extend its effective interval, "Ret. 1" will end at 10, and such ending will make "Ret. 2" at the first learning become invalid after 10 due to the Jenga assumption. So, we choose to review concept 1 at 9, and the effective interval of "Ret. 1" is thus extended to 26.

---

1 If there is no second group, i.e., reviewing, it will be shortened to $[4.1, 10]$ due to the Jenga assumption which means that the practical effective interval for concept 2 is also bounded by the ending moment of the effective interval for concept 1.

2 The reason for adopting a strategy whose learning duration is shorter is illustrated in message (2) above Section 6.3.2.

The "adaptive" paradigm means that we do not have to duplicate the original learning procedure (i.e., keep the concept reviewing sequence unchanged). As long as the precedent concepts are all active in an interval, such interval will support their following concepts' learning. Therefore, in order to take full advantage of "Ret. 2" interval earned at the first learning, a learner can directly review concept 3 in $[10.1, 11.1]$. The expected effective interval for concept 3 is thus extended from 11.1 to 27.1. However, since the interval of "Ret. 2" ends at 12.1, the maximum effective interval of the second group is $[11.1, 12.1]$, even if concept 2 is reviewed from 11.2 to 12.2. This is because the small blank slot $[12.1, 12.2]$ with concept 2's leave has already made concept 3 suspended, and we do not guarantee that concept 2's return at $[12.2, 28.2]$ can make concept 3 reactive without any extra effort. Therefore, in this circumstance, it seems that the "adaptive" paradigm is much worse than the "duplicating" paradigm in terms of the length of maximum intersecting effective interval (one time unit shown by the short thick line in Figure 6.9 vs. 13.8 time units calculated by (26 to 12.2) but without visualization in Figure 6.9).

However, if the spacing assumption can be relaxed, the conclusion will be completely different. If a learner is willing to give up the rest between concept 3 and concept 2's review (after all, in the original learning, the learner has a rest every two time units while in the reviewing, the learner can still have a rest every two time units, though the total times of rest are decreased), the maximum effective interval is $[11.1, 26]$ in Figure 6.10. Its length is 14.9 time units, which is more 1.1 time units than the 13.8 time units in the "duplicating" paradigm.

6.4.3 Implications

Figures 6.9 and 6.10 in this pilot study show that timing or scheduling plays an important role in maximizing the intersecting effective intervals. And the context of this clear contrast only allows one review strategy for all concepts rather than multiple strategy options for reviewing each concept. We can image that there will be probably myriads of changes when multiple strategy options for reviewing each concept are allowed. And the paradigm adopted in this pilot study is lazy and adaptive, not successive aggressive.

The second point is inside the second group of three "cabins," the original learning sequence for the three concepts is altered. To maximize the intersecting effective intervals, the positions of concept 2 and concept 3 are switched in their local cycle, i.e., the second time learning (reviewing), while obeying the Jenga assumption in a global perspective. Therefore, multi-space learning has now an additional connotation: spaces for learning not only can be superposed or decomposed but also can be switched between adjacent learning tasks in some circumstances. To maximize the effectiveness of multi-space learning, we still have lots of spaces to explore in the future.

6.5 CONNECTIONS TO MATROID THEORY: GREEDY ALGORITHM AND BEYOND

Theorem 6.1 shows that multi-space learning exhibits an optimal-substructure property in terms of maximizing intersecting effective intervals. Cormen et al. (2009, p. 442) provided proofs on "matroids exhibit the optimal-substructure property" Cormen et al. (2009, Lemma 16.10) and "correctness of the greedy algorithm on matroids" Cormen et al. (2009, Lemma 16.11).

This chapter actually focuses on scheduling. Cormen et al. (2009, p. 445) provided the proof that $(S, \mathcal{I})$ is a matroid where S is a set of unit-time tasks with deadlines and $\mathcal{I}$ is the set of all independent sets of tasks. Later, Cormen et al. (2022) enriched scheduling algorithms. While Brucker (2013) even published a monograph on the families of scheduling algorithms, which were organized in a unified format for readers to search solutions in different context as a dictionary. Other reference books on scheduling include (Leung, 2004; Yves & Frédéric, 2010). These monographs on scheduling are constrained in machines, e.g., micro-processors. The tasks scheduled usually have deadlines and designed penalties. To the best of our knowledge, these are the closest references to our study. However, the gap between the existing scheduling models or algorithms and our study is still huge. We have not found an existing scheduling model or algorithm can be applied to our study directly.

There are two reasons preventing us from translating existing scheduling algorithms into our solution. The first reason is the external objective: most of the existing scheduling algorithms aim at a minimum penalty. A minimum penalty implies that unexpected events can be tolerated as long as their summed penalties are small, or unexpected events do not prevent the whole tasks from completion even if some parts are overdue (suboptimal solutions exist). However, our study aims at maximizing the intersecting effective intervals. When some unexpected events happen, not only the maximizing mission will fail but also the whole task itself will face a failure risk. For example, if the duration of reviewing concept three is restored to two time units in Figure 6.10, it will waste the retention intervals contributed by concept 1's reviewing and concept 2's first time learning and fail the whole three-concept group's second retention.

The second reason is the internal mechanism: the context of existing scheduling models usually has an explicit deadline for each task. However, in our study, there is only an explicit sequence for concepts rather than explicit deadlines. Although a branch of existing scheduling algorithms is designed for dependent tasks, it does not have to face the situation that a "task" (concept) has to "grow up" (be learned/mastered) first and then "works" effectively (builds a base for later-coming concepts' learning). In other words, the "task" in our study has two parts, namely, the former part is learning and the latter part is taking effects. And the effective interval in our study is limited, unlike the general scheduling models where once a task is finished, its dependent tasks are always free to launch. That is to say, if a current scheduling model is capable to describe the learning-forgetting procedure of a human being, it has to maintain a two-track scheduling chart: one track is for the traditional life spans of tasks, and the other has to be designed for life spans of tasks' outcomes. And only during the life span of a task's outcome, can its dependent tasks be launched.

CHAPTER 7

Summary

MULTI-SPACE learning in our book is defined as a micro procedure of human learning which may take place in more than one space, with an implicit or explicit pursuit for effective learning strategies, and such effectiveness can be measured by the indicators based on the duration of knowledge retention. When a learner is in a space for learning or spaces for learning, we say this learner has chosen a learning strategy or a combination of multiple learning strategies to empower himself or herself. Therefore, the partition on spaces for learning is actually a partition on learning strategies.

In line with Bloom's taxonomy of educational objectives, an example of a partition on learning strategies is given. The subdividable structure of learning strategies enables learning strategies to be modeled as matroids. The primary goal of multi-space learning theory is to solve an optimizing problem, i.e., which combination of learning strategies can deliver the best effectiveness (in our theory, it is measured by the knowledge retention duration). Because learning strategies can be modeled as a matroid and matroids bring conveniences for optimization, matroid is regarded as the mathematical foundation of multi-space theory and as one corner stone added into the current foundations of learning sciences.

Currently, when we say the effectiveness of learning strategies, we usually think of meta-analysis studies on learning strategies, for example, meta-analysis on technology-enhanced learning (Betthäuser, Bach-Mortensen, & Engzell, 2023; Y. Lin & Yu, 2023; Xu, Zhao, Zhang, Liew, & Kogut, 2022; Sandy-Hanson, 2006), meta-analysis on vocabulary learning strategies (Yousefi & Biria, 2018; Mahdi, 2018; J. Lin & Lin, 2019; Yu & Trainin, 2022), and meta-analysis on math learning (Tokac, Novak, & Thompson, 2019;

DOI: 10.4324/9781032707013-7

Güler, Kokoc, & Önder Bütüner, 2023; Dondio, Gusev, & Rocha, 2023; Arztmann, Hornstra, Jeuring, & Kester, 2023).

Different from the meta-analysis on a specific learning strategy's effectiveness, we combine any specific learning strategy and the learning object it is applied into an abstract pair, $< object, strategy >$, and do not focus on a specific learning strategy's effectiveness at the stage of building multi-space learning theory. Adding parameters of learning intervals and effective intervals, such pair can be used to describe a micro learning procedure, which can be considered as a sequence of $< object/concept, strategy >$ pair. There are two major benefits of modeling learning strategies into a matroid: first of all, matroid can tolerate inconsistent empirical evidences on the effectiveness of a learning strategy or two learning strategies interfering with each other (see Section 5.2.3 as an example); second, it implies that once a learning strategy obtains its parameters' empirical values,[1] the whole algorithm or system on scheduling can thus run automatically to look for the optimal solution, as illustrated in Chapter 5 or Chapter 6.

From one fact (new learned knowledge can be forgotten after a while) and one principle (learning should be separated by spacing), we investigated matroid theory's applications in two domains, namely, vocabulary learning (Chapter 5) and math learning (Chapter 6). The difference between these two domains in the learning procedure is that math learning (well-structured) emphasizes the order of concepts to be learned, while vocabulary learning (ill-structured) does not require this prerequisite. In fact, these two chapters are examples on how to apply matroid theory to solve the effectiveness optimization problem in multi-space learning. Any interested reader can follow the similar paradigm to start their own explorations, including scratching from different ground truths or empirical evidences, e.g., in Tables 4.2 and 4.3, as we did which is mentioned at the beginning of this paragraph.

In Chapter 5, we initially assume that the effectiveness of two arbitrary learning strategies is linearly addable. Soon, the solved asymptotic retention rate (ARR) of a single learning strategy being negative leads us to speculate whether some learning strategies may interfere with each other. This is actually not a failure of matroid's application in ARR estimation, but

[1] The learning principles in Tables 4.2 and 4.3 are in fact summarized from multiple empirical studies. For example, FT(Feedback Timely) appeared in Table 5.1 is actually one of supporting evidences for feedback effects in Table 4.3.

a success of interference prediction by matroid theory from existing empirical data, and we indeed find the empirical evidences supporting such interference. To estimate single strategy's ARR more accurately, two matroid-related approaches are adopted: the "mean vertex" on a convex polyhedra, which is more explainable; and neural networks, which is more accurate. Therefore, when we pursuit the optimization combination of learning strategies, we have to keep in mind that due to the interference among strategies, the strategies combined are NOT "the more, the better."[2] In fact, we observed that the incremental effectiveness of adding new learning strategies into a smaller combination of strategies will be greater than adding them into a larger one, which makes the relationship between ARR and the number of combined strategies be a submodular function. In our context, the Gauss model is found to fit this relationship best. More details about the connections between our study and matroid theory can be found in the subsections inside this chapter. Our extended work (Li & Wang, 2023) on identifying learning strategies based on massive user data in a language learning application is a re-examination on our theory. It validates that the best learning strategy can bring an optimal learning return on investment, something that users may not even be aware of.

Besides estimating the single strategy's ARR and optimizing combined strategies' ARR in Chapter 5, two scheduling optimization problems are studied in Chapter 6. They both are on maximizing the intersecting effective interval, but the first one is in the first-time learning context, while the second one is in the context of reviewing, i.e., the second-time learning. Specifically, the first problem is: if there are two learning strategies for a learner to choose, one strategy (A) has a longer effective duration and a longer learning duration than the other (B), which strategy should this learner choose? It is probably obvious that this learner should choose A because A has a longer effective duration; however, if this learner chooses B, he or she can earn an earlier effective time than A, so that this learner may also accumulate a longer intersecting effective interval for following concepts. We examined this problem and proved that by a greedy algorithm, the global maximum intersecting effective interval consists of local maximum intersecting effective intervals. Therefore, for the first-time learning in our designed example, the solution to the optimal schedule which has the maximum intersecting effective interval is all the concepts

[2] A Hollywood film entitled "The Forbidden Kingdom" exhibited a funny plot that two trainers cannot teach a trainee well.

ought to be learned with a learning strategy whose effective duration is the longest except the last one, and the last one ought to be learned with a learning strategy whose learning duration is the shortest.

As aforementioned, the difference in terms of prerequisite between Chapter 5 and Chapter 6 is that the latter highlights the order of concepts to be learned. Such prerequisite requires us to take care of affected concepts when their precedent concept is forgotten, rather than treat them as context-free, independent concepts. Therefore, we assume that if a precedent concept is forgotten, all of its following concepts have to be reviewed (learning at the second time), and this is also the reason for the first problem's formulation. As new learned knowledge can be forgotten after a while, we need to review. Our study shows that the timing of review is also vital in the scheduling optimization problem which maximizes the intersecting effective interval. More importantly, we believe this is worth exploring in the future, since no on-the-shelf algorithm can be directly applied to this scenario: one task can only be implemented during the effective interval of its precedent task, and their maximum intersecting effective interval is the optimal target. More details about the gap between existing scheduling algorithms and the scenario we face is discussed in Section 6.5. In the future, as the number of related empirical evidences (e.g., (Rohrer & Taylor, 2006; Hopkins, Lyle, Hieb, & Ralston, 2016; Lyle, Bego, Hopkins, Hieb, & Ralston, 2020; Yang & Chen, 2021; Walsh et al., 2023)) increases, more useful parameters[3] that fit our context will be found as we did in Chapter 5, so that this theory can be further examined by more practical data.

As we write here, we are about to conclude this manuscript. Though it feels more like we are just getting started, we find it challenging to end. "Nothing is final (in science)." Recently, Matayoshi and Uzun (2022) have reported their contributions on incorporating forgetting within a knowledge structure, and they have also found that satisfying positive knowledge correlation implies that the knowledge structure is closed under both union and intersection. We believe their contributions will shed light on our future work. However, we have to pause here as our staged work. We are looking forward to any comments on our manuscript to push this work forward together.

[3] Currently, these parameters do not focus on the scenario where a forgotten concept will affect a concept to be mastered.

References

Aggarwal, C. (Ed.). (2014). *Data classification: Algorithms and applications.* New York, NY: Taylor & Francis Group. doi: 10.1201/b17320

Akaike, H. (1974). A new look at the statistical model identification. *IEEE Transactions on Automatic Control, 19*(6), 716–723. doi: 10.1109/TAC.1974.1100705

Alammary, A., Sheard, J., & Carbone, A. (2014). Blended learning in higher education: Three different design approaches. *Australasian Journal of Educational Technology, 30*(4), 440–454. doi: 10.14742/ajet.693

Allen, C., DiGiacomo, D., Van Horne, K., & Penuel, W. (2018). Pursuing interests and getting involved: Exploring the conditions of sponsorship in youth learning. *Digital Education Review, 33*, 120–129. (https://revistes.ub.edu/index.php/der/article/view/21809, Last access on 2023-06-09)

Almond, R., Mislevy, R., Steinberg, L., Yan, D., & Williamson, D. (2015). *Bayesian networks in educational assessment.* New York, NY: Springer. doi: 10.1007/978-1-4939-2125-6

Alshehri, E., Alhakami, H., Baz, A., & Alsubait, T. (2020). A comparison of edm tools and techniques. *International Journal of Advanced Computer Science and Applications, 11*(12), 824–831. doi: 10.14569/IJACSA.2020.0111295

Amarasinghe, I., Hernández-Leo, D., & Jonsson, A. (2019). Data-informed design parameters for adaptive collaborative scripting in across-spaces learning situations. *User Modeling and User-Adapted Interaction, 29*(4), 869–892. doi: 10.1007/s11257-019-09233-8

Anderson, J. R. (1996). Act: A simple theory of complex cognition. *American Psychologist, 51*(4), 355–365. doi: 10.1037/0003-066X.51.4.355

Anderson, L. W., & Krathwohl, D. R. (Eds.). (2001). *A taxonomy for learning, teaching, and assessing: A revision of bloom's taxonomy of educational objectives.* New York: Pearson, Allyn & Bacon.

Andler, D. (2012). Mathematics in cognitive science. In M. Weber, D. Dieks, W. Gonzalez, S. Hartmann, & M. Stöltzner (Eds.), *Probabilities, laws, and structures* (pp. 363–377). Dordrecht: Springer.

Arztmann, M., Hornstra, L., Jeuring, J., & Kester, L. (2023). Effects of games in stem education: a meta-analysis on the moderating role of student background characteristics. *Studies in Science Education, 59*(1), 109–145. doi: 10.1080/03057267.2022.2057732

Ashby, G., Colonius, H., & Dzhafarov, E. (Eds.). (2023). *New handbook of mathematical psychology: Volume 3, perceptual and cognitive processes.* Cambridge, Massachusetts: Cambridge University Press.

Averell, L., & Heathcote, A. (2011). The form of the forgetting curve and the fate of memories. *Journal of Mathematical Psychology, 55*(2011), 25–35. doi: 10.1016/j.jmp.2010.08.009

Azcona, D., Hsiao, I.-H., & Smeaton, A. (2019). Detecting students-at-risk in computer programming classes with learning analytics from students' digital footprints. *User Modeling and User-Adapted Interaction, 29*(4), 759–788. doi: 10.1007/s11257-019-09234-7

Baddeley, A. (1992). Working memory. *Science, 255*(5044), 556–559. doi: 10.1126/science.1736359

Bakker, A., Demerouti, E., Oerlemans, W., & Sonnentag, S. (2013). Workaholism and daily recovery: A day reconstruction study of leisure activities. *Journal of Organizational Behavior, 34*(1), 87–107. doi: 10.1002/job.1796

Batchelder, W., Colonius, H., & Dzhafarov, E. (Eds.). (2021). *New handbook of mathematical psychology: Volume 2, modeling and measurement.* Cambridge, Massachusetts: Cambridge University Press.

Batchelder, W., Colonius, H., Dzhafarov, E., & Myung, J. (Eds.). (2016). *New handbook of mathematical psychology: Volume 1, foundations and methodology.* Cambridge, Massachusetts: Cambridge University Press.

Bazgana, C., Toubaline, S., & Tuza, Z. (2011). The most vital nodes with respect to independent set and vertex cover. *Discrete Applied Mathematics, 159*, 1933–1946. doi: 10.1016/j.dam.2011.06.023

Berg, A., Scheffel, M., Drachsler, H., Ternier, S., & Specht, M. (2016). The dutch xapi experience. In D. Gašević & G. Lynch (Eds.), *Proceedings of the sixth international conference on learning analytics & knowledge* (pp. 544–545). New York, NY, United States: Association for Computing Machinery. doi: 10.1145/2883851.2883968

Betthäuser, B., Bach-Mortensen, A., & Engzell, P. (2023). A systematic review and meta-analysis of the evidence on learning during the covid-19 pandemic. *Nature Human Behaviour*, *7*, 375–385. doi: 10.1038/s41562-022-01506-4

Bonifay, W. (2020). *Multidimensional item response theory*. Thousand Oaks, California, USA: Sage Publications, Inc.

Bowman, D., Swiecki, Z., Cai, Z., Wang, Y., Eagan, B., Linderoth, J., & Shaffer, D. W. (2021). The mathematical foundations of epistemic network analysis. In A. Ruis & S. Lee (Eds.), *Advances in quantitative ethnography* (pp. 91–105). Cham: Springer International Publishing. doi: 10.1007/978-3-030-67788-6_7

Breiman, L. (2001). Random forests. *Machine Learning*, *45*, 5–32. doi: 10.1023/A:1010933404324

Broadbent, J., & Poon, W. (2015). Self-regulated learning strategies & academic achievement in online higher education learning environments: A systematic review. *Internet and Higher Education*, *27*, 1–13. doi: 10.1016/j.iheduc.2015.04.007

Brown, M. (2005). Learning spaces. In D. Oblinger & J. Oblinger (Eds.), *Educating the net generation* (pp. 174–195). Denver, CO, United States: Educause.

Brucker, P. (2013). *Scheduling algorithms (4th ed.)*. Berlin, Heidelberg: Springer. doi: 10.1007/978-3-540-24804-0

Burnham, K., & Anderson, D. (2002). *Model selection and multimodel inference, a practical information-theoretic approach (2nd ed.)*. Springer-Verlag New York, Inc.

Cardona, T., & Cudney, E. (2019). Predicting student retention using support vector machines. *Procedia Manufacturing*, *39*, 1827–1833. doi: https://doi.org/10.1016/j.promfg.2020.01.256

Caruana, R., & Niculescu-Mizil, A. (2006). An empirical comparison of supervised learning algorithms. In *Proceedings of the 23rd international conference on machine learning* (pp. 161–168). New York, NY, USA. doi: 10.1145/1143844.1143865

Castells, N. M., & Illera, J. L. R. (2018). Activity and learning contexts in educational transmedia. *Digital Education Review*, *33*, 77–86. (https://revistes.ub.edu/index.php/der/article/view/21704, Last access on 2023-06-09)

Catrysse, L., Gijbels, D., Donche, V., De Maeyer, S., Lesterhuis, M., & Van den Bossche, P. (2018). How are learning strategies reflected in the eyes? combining results from self-reports and eye-tracking. *British Journal of Educational Psychology*, *88*(1), 118–137. doi: 10.1111/bjep.12181

Charitopoulos, A., Rangoussi, M., & Koulouriotis, D. (2020). On the use of soft computing methods in educational data mining and learning analytics research: a review of years 2010–2018. *International Journal of Artificial Intelligence in Education*, *30*, 371–-430. doi: 10.1007/s40593-020-00200-8

Chekuri, C. (2018, Spring). *Packing problems: maximum independent set, packing integer programs.* Approximation Algorithms (CS 583) Lecture 5. (https://courses.engr.illinois.edu/cs583/sp2018/Notes/packing.pdf)

Chen, C. M., & Hsu, S. H. (2008). Personalized intelligent mobile learning system for supporting effective english learning. *Educational Technology & Society*, *11*(3), 153–180. (https://www.jstor.org/stable/jeductechsoci.11.3.153)

Conley, A. M. (2012). Patterns of motivation beliefs: Combining achievement goal and expectancy-value perspectives. *Journal of Educational Psychology*, *104*(1), 32–47. doi: 10.1037/a0026042

Cormen, T., Leiserson, C., Rivest, R., & Stein, C. (2009). *Introduction to algorithms (3rd ed.).* Cambridge, Massachusetts: The MIT Press.

Cormen, T., Leiserson, C., Rivest, R., & Stein, C. (2022). *Introduction to algorithms (4th ed.).* Cambridge, Massachusetts: The MIT Press.

Cortés, S., García-Pernía, M. R., Fuente, J., Martínez-Borda, R., & Lacasa, P. (2018). Young creators in open spaces: Digital ethnography. *Digital Education Review*, *33*, 185–202. (https://revistes.ub.edu/index.php/der/article/view/21623, Last access on 2023-06-09)

Cosyn, E., & Uzun, H. (2009). Note on two necessary and sufficient axioms for a well-graded knowledge space. *Journal of Mathematical Psychology*, *53*, 40–42. doi: 10.1016/j.jmp.2008.09.005

Cosyn, E., Uzun, H., Doble, C., & Matayoshi, J. (2021). A practical perspective on knowledge space theory: Aleks and its data. *Journal of Mathematical Psychology*, *101*(6), 102512. doi: 10.1016/j.jmp.2021.102512

Cropley, M., Weidenstedt, L., Leick, B., & Sütterline, S. (2022). Working from home during lockdown: the association between rest breaks and well-being. *Ergonomics*, *0*(0), 1–11. doi: 10.1080/00140139.2022.2095038

Culbertson, M. (2016). Bayesian networks in educational assessment: The state of the field. *Applied psychological measurement*, *40*(1), 3–21. doi: 10.1177/0146621615590401

Cutler, A., Cutler, D., & Stevens, J. (2012). Random forests. In C. Zhang & Y. Ma (Eds.), *Ensemble machine learning: Methods and applications* (pp. 157–175). New York, NY: Springer. doi: 10.1007/978-1-4419-9326-7_5

Dantzig, G., Orden, A., & Wolfe, P. (1955). Generalized simplex method for minimizing a linear form under linear inequality restraints. *Pacific Journal Mathematics*, *5*(2), 183–195.

Davis, D., Chen, G., Hauff, C., & Houben, G.-J. (2018). Activating learning at scale: A review of innovations in online learning strategies. *Computers & Education*, *125*, 327–344. doi: 10.1016/j.compedu.2018.05.019

Dietrich, B. L. (1989). Matroids and antimatroids - a survey. *Discrete Mathematics*, *78*(3), 223–237. doi: 10.1016/0012-365X(89)90180-5

Doignon, J., & Falmagne, J. (2016). Knowledge spaces and learning spaces. In W. Batchelder, H. Colonius, E. Dzhafarov, & J. Myung (Eds.), *New handbook of mathematical psychology (cambridge handbooks in psychology)* (pp. 274–321). Cambridge: Cambridge University Press. doi: 10.1017/9781139245913.006

Dondio, P., Gusev, V., & Rocha, M. (2023). Do games reduce maths anxiety? a meta-analysis. *Computers & Education*, *194*, 104650. doi: 10.1016/j.compedu.2022.104650

Dowson, M., & McInerney, D. M. (2004). The development and validation of the goal orientation and learning strategies survey (goals-s). *Educational and Psychological Measurement*, *64*(2), 290–310. doi: 10.1177/0013164403251335

Dweck, C. S. (2006). *Mindset: The new psychology of success.* Random House. New York, USA.

Ebbinghaus, H. (1885/1913). *Memory: A contribution to experimental psychology.* (H.A. Ruger & C.E. Bussenius, trans.) Teachers College, Columbia University. New York, USA.

El-Bishouty, M. M., Ogata, H., & Yano, Y. (2007). Perkam: personalized knowledge awareness map for computer supported ubiquitous learning. *Educational Technology & Society*, *10*(3), 122–134. (http://www.jstor.org/stable/jeductechsoci.10.3.122)

Engel, A., Coll, C., Membrive, A., & Oller, J. (2018). Information and communication technologies and students' out-of-school learning experiences. *Digital Education Review, 33*, 130–149. (https://revistes.ub.edu/index.php/der/article/view/21689, Last access on 2023-06-09)

Erten, I., & Tekin, M. (2008). Effects on vocabulary acquisition of presenting new words in semantic sets versus semantically unrelated sets. *System, 36*, 407–422. doi: 10.1016/j.system.2008.02.005

Esteban-Guitart, M., Coll, C., & Penuel, W. (2018). Learning across settings and time in the digital age. *Digital Education Review, 33*. (https://revistes.ub.edu/index.php/der/issue/view/1830, Last access on 2023-06-09)

Falmagne, J.-C., & Doignon, J.-P. (2011). *Learning spaces: Interdisciplinary applied mathematics*. Springer Berlin, Heidelberg. doi: 10.1007/978-3-642-01039-2

Feige, U., Mirrokni, V. S., & Vondrak, J. (2011). Maximizing non-monotone submodular functions. *Society for Industrial and Applied Mathematics(SIAM) Journal on Computing, 40*(4), 1133–1153. doi: 10.1137/090779346

Finkbeiner, M., & Nicol, J. (2003). Semantic category effects in second language word learning. *Applied Psycholinguistics, 24*, 369–383. doi: 10.1017/S0142716403000195

Fischer, F., Kollar, I., Ufer, S., Sodian, B., & J., E. (2014). Scientific reasoning and argumentation: Advancing an interdisciplinary research agenda in education. *Frontline Learning Research, 2*(3), 28–45. doi: 10.14786/flr.v2i3.96

García-Pastor, M. D. (2018). Learner identity in efl: An analysis of digital texts of identity in higher education. *Digital Education Review, 33*, 55–76. (https://revistes.ub.edu/index.php/der/article/view/21632, Last access on 2023-06-09)

Güler, M., Kokoc, M., & Önder Bütüner, S. (2023). Does a flipped classroom model work in mathematics education? a meta-analysis. *Education and Information Technologies, 28*(3), 57–79. doi: 10.1007/s10639-022-11143-z

Geary, D. (2021). *Preparation for success in algebra: Exploring math education relationships by analyzing large data sets (emeralds) final report 2021* (Tech. Rep.). Pittsburgh, PA: Human-Computer Interaction Institute at Carnegie Mellon University and Educational Testing Service. (https://achievethecore.org/page/

3367/preparation-for-success-in-algebra-exploring-math-education-relationships-by-analyzing-large-data-sets-emeralds, Last access on 2023-08-31)

Glogger, I., Schwonke, R., Holzäpfel, L., Nückles, M., & Renkl, A. (2012). Learning strategies assessed by journal writing: Prediction of learning outcomes by quantity, quality, and combinations of learning strategies. *Journal of Educational Psychology, 104*(2), 452–468. doi: 10.1037/a0026683

González-Patiño, J. (2018). Learn, teach and share, participation in expanded educational communities: Madrid pikler seminar as a practice of continuing education between childhood professionals. *Digital Education Review, 33*, 203–216. (https://revistes.ub.edu/index.php/der/article/view/21630, Last access on 2023-06-09)

Gossen, H. (1854). *The laws of human relations and the rules of human action derived therefrom (1983).* Cambridge, MA: MIT Press.

Graesser, A. C. (2009). Inaugural editorial for journal of educational psychology. *Journal of Educational Psychology, 101*(2), 259–261. doi: 10.1037/a0014883

Hagan, M., Demuth, H., Beale, M., & De Jesús, O. (2014). *Neural network design (2nd edition).* Martin Hagan. Oklahoma State University, Stillwater, Oklahoma, US.

Hambleton, R., & Swaminathan, H. (1985). *Item response theory: Principles and applications.* New York, USA: Springer Science+Business Media, LLC.

Hao, T., Wang, Z., & Ardasheva, Y. (2021). Technology-assisted vocabulary learning for efl learners: A meta-analysis. *Journal of Research on Educational Effectiveness, 14*(3), 645–667. doi: 10.1080/19345747.2021.1917028

Hernández-Hernández, F., Sancho-Gil, J., & Domingo-Coscollola, M. (2018). Cartographies as spaces of inquiry to explore of teachers' nomadic learning trajectories. *Digital Education Review, 33*, 105–119. (https://revistes.ub.edu/index.php/der/article/view/21622, Last access on 2023-06-09)

Hernandez, M., Johnson, A., & Werkenthin, A. (2020). Understanding and applying cmi5 in an xapi solution [video presentation]. In *Proceedings of the 2020 interservice/industry training, simulation, and education conference (i/itsec).* Arlington,

VA, United States: National Training and Simulation Association. Retrieved from https://adlnet.gov/videos/2021/05/20210503-understanding-and-applying-cmi5-in-an-xapi-solution/

Higa, M. (1963). Interference effects of intralist word relationships in verbal learning. *Journal of Verbal Learning and Verbal Behavior*, *2*(2), 170–175. doi: 10.1016/S0022-5371(63)80082-1

Hofmann, J. (2006). Why blended learning hasn't (yet) fulfilled its promises: Answers to those questions that keep you up at night. In C. Bonk & C. Graham (Eds.), *Handbook of blended learning: Global perspectives, local designs* (pp. 27–40). San Francisco, CA, United States: Pfeiffer Publishing.

Hoover, C., Ragsdale, J., & Ayres, T. (2022). An experimental test of resource recovery from physical and relaxation work break activities. *Stress & Health*, *38*(3), 477–489. doi: 10.1002/smi.3108

Hopkins, R., Lyle, K., Hieb, J., & Ralston, P. (2016). Spaced retrieval practice increases college students' short- and long-term retention of mathematics knowledge. *Educational Psychology Review*, *28*, 853–873. doi: 10.1007/s10648-015-9349-8

Hunter, E., & Wu, C. (2016). Give me a better break: Choosing workday break activities to maximize resource recovery. *Journal of Applied Psychology*, *101*(2), 302–311. doi: 10.1037/apl0000045

Hutt, S., Krasich, K., Mills, C., Bosch, N., White, S., Brockmole, J., & D' Mello, S. (2019). Automated gaze-based mind wandering detection during computerized learning in classrooms. *User Modeling and User-Adapted Interaction*, *29*(4), 821–867. doi: 10.1007/s11257-019-09228-5

Irfan, M., & Gudivada, V. (2016). Cognitive computing applications in education and learning. *Handbook of Statistics*, *35*, 283–300. doi: 10.1016/bs.host.2016.07.008

Issah, I., Appiah, O., Appiahene, P., & Inusah, F. (2023). A systematic review of the literature on machine learning application of determining the attributes influencing academic performance. *Decision Analytics Journal*, *7*, 100204. doi: 10.1016/j.dajour.2023.100204

Jeng, Y.-L., Wu, T.-T., Huang, Y.-M., Tan, Q., & Yang, S. J. H. (2010). The add-on impact of mobile applications in learning strategies: A review study. *Educational Technology & Society*, *13*(3), 3–11. (https://www.jstor.org/stable/jeductechsoci.13.3.3)

Jenkins, J., Pany, D., & Schreck, J. (1978). *Vocabulary and reading comprehension: Instructional effects (technical report no. 100)* (Tech. Rep.). Urbana: University of Illinois, Center for the Study of Reading.

Jiang, J., Liu, L., Zong, P., Tao, F., Lai, H., Chen, Y., & Mi, C. (2022). What makes online-merge-offline (omo) education succeed? a holistic success model. In G. Meiselwitz et al. (Eds.), *Hci international 2022 - late breaking papers. interaction in new media, learning and games* (pp. 297–308). Cham: Springer Nature Switzerland. doi: 10.1007/978-3-031-22131-6_23

Jonathans, P., Widiati, U., Astutik, I., & Ratri, D. (2021). The practices of intentional vocabulary acquisition for asian efl learners: A systematic review. *English Review: Journal of English Education*, *9*(2), 253–261. (https://journal.uniku.ac.id/index.php/ERJEE/article/view/4350/2560, Last access on 2023-01-25)

Jornet, A., & Erstad, O. (2018). From learning contexts to learning lives: Studying learning (dis)continuities from the perspective of the learners. *Digital Education Review*, *33*, 1–25. (https://revistes.ub.edu/index.php/der/article/view/21810, Last access on 2023-06-09)

Kalyuga, S. (2011). Cognitive load theory: how many types of load does it really need? *Educational Psychology Review*, *23*, 1–19. doi: 10.1007/s10648-010-9150-7

Kamada, T., & Kawai, S. (1989). An algorithm for drawing general undirected graphs. *Information Processing Letters*, *31*(1), 7--15. doi: 10.1016/0020-0190(89)90102-6

Kaur, A., Bhatia, M., & Stea, G. (2022). A survey of smart classroom literature. *Teaching in Higher Education*, *12*(2), 86. doi: 10.3390/educsci12020086

Kingston, N., & Dorans, N. J. (1982). *The feasibility of using item response theory as a psychometric model for the gre aptitude test* (Report No. RR-82-12). English Testing Service. doi: 10.1002/j.2333-8504.1982.tb01298.x

Kirschner, P., Sweller, J., Kirschner, F., & Zambrano, J. R. (2018). From cognitive load theory to collaborative cognitive load theory. *International Journal of Computer-Supported Collaborative Learning*, *13*, 213–233. doi: 10.1007/s11412-018-9277-y

Korte, B., & Lovász, L. (1988). The intersection of matroids and antimatroids. *Discrete Mathematics*, *73*(1-2), 143–157. doi: 10.1016/0012-365X(88)90142-2

Korte, B., Lovász, L., & Schrader, R. (1991). Greedoids. In R. Graham, M. Hill, B. Korte, & L. Lovász (Eds.), *Algorithms and combinatorics* (Vol. 4, p. 41). Berlin: Springer.

Kortsch, K., Schulte, E., & Kauffeld, S. (2019). Learning work: informal learning strategies of german craft workers. *European Journal of Training and Development*, *43*(5/6), 418–434. doi: 10.1108/EJTD-06-2018-0052

Kreinovich, V. (1991). Arbitrary nonlinearity is sufficient to represent all functions by neural networks: A theorem. *Neural Networks*, *4*, 381–383. doi: 10.1016/0893-6080(91)90074-F

Kruskal, J. (1956). On the shortest spanning subtree of a graph and the traveling salesman problem. *Proceedings of the American Mathematical Society*, *7*(1), 48–50. doi: 10.1090/S0002-9939-1956-0078686-7

Kumpulainen, K., Kajamaa, A., & Rajala, A. (2018). Understanding educational change: Agency-structure dynamics in a novel design and making environment. *Digital Education Review*, *33*, 26–38. (https://revistes.ub.edu/index.php/der/article/view/21633, Last access on 2023-06-09)

Kwet, M., & Prinsloo, P. (2020). The 'smart' classroom: a new frontier in the age of the smart university. *Teaching in Higher Education*, *25*(4), 510–526. doi: 10.1080/13562517.2020.1734922

Lai, M., Lam, K., & Lim, C. (2016). Design principles for the blend in blended learning: A collective case study. *Teaching in Higher Education*, *21*(6), 716–729. doi: 10.1080/13562517.2016.1183611

Lalor, J., Wu, H., & Yu, H. (2016). Building an evaluation scale using item response theory. In *Proceedings of the conference on empirical methods in natural language processing* (pp. 648–657). doi: 10.18653/v1/d16-1062

Lee, W., Bartlett, P., & Williamson, R. (1996). Efficient agnostic learning of neural networks with bounded fan-in. *IEEE Transactions on Information Theory*, *42*(6), 2118–2132. doi: 10.1109/18.556601

Leena, H. A., Ranim, S. A., Mona, S. A., Dana, K. A., Irfan, U. K., & Nida, A. (2021). Predicting student academic performance using support vector machine and random forest. In *Proceedings of the 2020 3rd international conference on education technology management* (pp. 100–107). New York, NY, USA: Association for Computing Machinery. doi: 10.1145/3446590.3446607

Leung, J. Y.-T. (Ed.). (2004). *Handbook of scheduling: Algorithms, models, and performance analysis.* Boca Raton, Florida, USA: Taylor & Fancis Group. doi: 10.1201/9780203489802

Levinson, A., & Barron, B. (2018). Latino immigrant families learning with digital media across settings and generations. *Digital Education Review, 33*, 150–169. (https://revistes.ub.edu/index.php/der/article/view/21813, Last access on 2023-06-09)

Li, M., & Wang, T. (2023). Optimizing learning return on investment: Identifying learning strategies based on user behavior characteristic in language learning applications. *Education and Information Technologies.* doi: 10.1007/s10639-023-12078-9

Limeri, L., Carter, N., Choe, J., Harper, H., Martin, H., Benton, A., & Dolan, E. (2020). Growing a growth mindset: characterizing how and why undergraduate students' mindsets change. *International Journal of STEM Education, 7*(35). doi: 10.1186/s40594-020-00227-2

Lin, J., & Lin, H. (2019). Mobile-assisted esl/efl vocabulary learning: A systematic review and meta-analysis. *Computer Assisted Language Learning, 32*(8), 878–919. doi: 10.1080/09588221.2018.1541359

Lin, Y., & Yu, Z. (2023). A meta-analysis of the effects of augmented reality technologies in interactive learning environments(2012-2022). *Computer Applications in Engineering Education, 31*(4), 1111–1131. doi: 10.1002/cae.22628

Liu, Y., Wang, T., Bo, H., & Zhang, N. (2022). The influence of personality on epistemic network in an open-ended question discussion scenario. In H. Ma & Z. Niu (Eds.), *4th international conference on computer science and technologies in education (cste)* (pp. 215–220). Piscataway, NJ, USA: IEEE Computer Society. doi: 10.1109/CSTE55932.2022.00046

Lozin, V. (2017). From matchings to independent sets. *Discrete Applied Mathematics, 231*, 4–14. doi: 10.1016/j.dam.2016.04.012

Luce, R. (1964). The mathematics used in mathematical psychology. *The American Mathematical Monthly, 71*(4), 364-378. doi: 10.1080/00029890.1964.11992249

Lyle, K., Bego, C., Hopkins, R., Hieb, J., & Ralston, P. (2020). How the amount and spacing of retrieval practice affect the short- and long-term retention of mathematics knowledge. *Educational Psychology Review, 32*, 277–295. doi: 10.1007/s10648-019-09489-x

Ma, R. (2022). *Top english learning mobile apps in china.* https://www.classcentral.com/report/china-english-learning-apps/. (Accessed: 2023-02-22)

Mahdi, H. (2018). Effectiveness of mobile devices on vocabulary learning: A meta-analysis. *Journal of Educational Computing Research, 56*(1), 134–154. doi: 10.1177/0735633117698826

Mahmood, U. H., & Miller, F. (2022). Discrimination with unidimensional and multidimensional item response theory models for educational data. *Communications in Statistics - Simulation and Computation, 51*(6), 2992-3012. doi: 10.1080/03610918.2019.1705344

Manso-Vázquez, M., Caeiro-Rodríguez, M., & Llamas-Nistal, M. (2018). An xapi application profile to monitor self-regulated learning strategies. *IEEE Access, 6*, 42467 – 42481. doi: 10.1109/ACCESS.2018.2860519

Martinez-Maldonado, R., Hernández-Leo, D., & Pardo, A. (2019). Preface to the special issue on learning analytics and personalized support across spaces. *User Modeling and User-Adapted Interaction, 29*(4), 751–758. doi: 10.1007/s11257-019-09243-6

Matayoshi, J., & Uzun, H. (2022). Learning, forgetting, and the correlation of knowledge in knowledge space theory. *Journal of Mathematical Psychology, 109*, 102674. doi: 10.1016/j.jmp.2022.102674

Matt, J. (2021). *Analyze n-dimensional convex polyhedra.* https://www.mathworks.com/matlabcentral/fileexchange/30892-analyze-n-dimensional-convex-polyhedra. (Accessed: 2023-02-22)

Matzavela, V., & Alepis, E. (2021). Decision tree learning through a predictive model for student academic performance in intelligent m-learning environments. *Computers and Education: Artificial Intelligence, 2*, 100035. doi: 10.1016/j.caeai.2021.100035

Mayer, R. (1988). Learning strategies: An overview. In C. Weinstein, E. Goetz, & P. Alexander (Eds.), *Learning and study strategies: Issues in assessment, instruction, and evaluation* (pp. 11–22). New York: Academic Press.

Mayer, R. (2021). Evidence-based principles for how to design effective instructional videos. *Journal of Applied Research in Memory and Cognition.* doi: 10.1016/j.jarmac.2021.03.007

McCombs, B. (1988). Motivational skills training: Combining metacognitive, cognitive, and affective learning strategies. In C. Weinstein, E. Goetz, & P. Alexander (Eds.), *Learning and study strategies: Issues in assessment, instruction, and evaluation* (pp. 141–168). New York: Academic Press.

McConnell, D., Chapman, L., Czajka, D. C., Jones, J., Ryker, K. D., & J., W. (2017). Instructional utility and learning efficacy of common active learning strategies. *Journal of Geoscience Education, 65*(4), 604–625. doi: 10.5408/17-249.1

Moody, S., Hu, X., Xu, Z., & Lee, S. (2018). Vocabulary instruction: A critical analysis of theories, research and practice. *Education Sciences, 8*(180). doi: 10.3390/educsci8040180

Moore, M. (1973). Toward a theory of independent learning and teaching. *The Journal of Higher Education, 44*(9), 661–679. doi: 10.1080/00221546.1973.11776906

Moore, M. (1991). Editorial: Distance education theory. *The American Journal of Distance Education, 5*(3), 1–5. doi: 10.1080/08923649109526758

Mueller, P. A., & Oppenheimer, D. M. (2014). The pen is mightier than the keyboard: Advantages of longhand over laptop note taking. *Psychological Science, 25*(6), 1159–1168. doi: 10.1177/0956797614524581

Murre, J., & Chessa, A. (2022). Why ebbinghaus' savings method from 1885 is a very 'pure' measure of memory performance. *Psychonomic Bulletin & Review*. doi: 10.3758/s13423-022-02172-3

Murre, J., Chessa, A., & Metter, M. (2013). A mathematical model of forgetting and amnesia. *Frontiers in Psychology, 4*(76). doi: 10.3389/fpsyg.2013.00076

Murre, J., & Dros, J. (2014). Replication and analysis of ebbinghaus' forgetting curve. *PLoS ONE, 10*(7). doi: 10.1371/journal.pone.0120644

Murre, J., Meeter, M., & Chessa, A. (2007). Modeling amnesia: Connectionist and mathematical approaches. In M. Wenger & C. Schuster (Eds.), *Statistical and process models for neuroscience and aging* (pp. 119–162). Mahwah, NJ: Lawrence Eribaum.

Narayanan, H. (1997). *Submodular functions and electrical networks.* Amsterdam; New York: Elsevier.

Nash, H., & Snowling, M. (2006). Teaching new words to children with poor existing vocabulary knowledge: A controlled evaluation of the definition and context methods. *International Journal of Language & Communication Disorders, 41*(3), 335–354. doi: 10.1080/13682820600602295

Nation, K., Snowling, M., & Clarke, P. (2007). Dissecting the relationship between language skills and learning to read: Semantic and phonological contributions to new vocabulary learning in children with poor reading comprehension. *Advances in Speech–Language Pathology, 9*(2), 131–139. doi: 10.1080/14417040601145166

NICHHD. (2000). *Report of the national reading panel. teaching children to read: An evidence-based assessment of the scientific research literature on reading and its implications for reading instruction (nih publication no. 00-4769)* (Tech. Rep.). Washingtong, D.C., U.S. Goverment Printing Office: National Institute of Child Health and Human Development.

Noble, W. (2006). What is a support vector machine? *Nature Biotechnology, 24*, 1565–1567. doi: 10.1038/nbt1206-1565

NRTAC. (2010). *A review of the current research on vocabulary instruction* (Tech. Rep.). RMC Research Corporation, Portsmouth, NH, USA: National Reading Technical Assistance Center.

Overton, T., & Johnson, L. (2016, July). *Evidence-based practice in learning and teaching for stem disciplines.* Australian Council of Deans of Science. (http://www.acds-tlcc.edu.au/wp-content/uploads/sites/14/2016/07/ACDS-stem-principles-WEB.pdf)

Oxford, R. (1990). *Language learning strategies: What every teacher should know.* New York: Newbury House.

Oxley, J. (2001). On the interplay between graphs and matroids. In J. Hirschfeld (Ed.), *Surveys in combinatorics (london math. soc. lecture notes 288)* (p. 199-239). Cambridge, UK: Cambridge University Press.

Oxley, J. G. (1992). *Matroid theory.* New York: Oxford University Press.

Paria, B., Poczos, B., Ravikumar, P., Schneider, J., & Suggala, A. S. (2022). Be greedy –a simple algorithm for blackbox optimization using neural networks. In *International conference on machine learning 2022 workshop on adaptive experimental design and active learning in the real world* (pp. 1–27). Baltimore, United States.

Patricia, A., Enrico, V., Shariq, B., De la Hoz Franco, E., Alberto, P., … Fulvio, P. (2022). Machine learning applied to datasets of human activity recognition: Data analysis in health care. *Current Medical Imaging, 19*(1), 46–64. doi: 10.2174/1573405618666220104114814

Pelánek, R. (2017). Bayesian knowledge tracing, logistic models, and beyond: an overview of learner modeling techniques. *User Modeling and User-Adapted Interaction*, *27*, 313–350. doi: 10.1007/s11257-017-9193-2

Peltokorpi, J., & Jaber, M. (2022). Interference-adjusted power learning curve model with forgetting. *International Journal of Industrial Ergonomics*, *88*. doi: 10.1016/j.ergon.2021.103257

Perez, T., Cromley, J. G., & Kaplan, A. (2014). The role of identity development, values, and costs in college stem retention. *Journal of Educational Psychology*, *106*(1), 315–329. doi: 10.1037/a0034027

Plaskura, P. (2019). Modelling of forgetting curves in educational e-environment. *Information Technologies and Learning Tools*, *71*(3), 1–11. doi: 10.33407/itlt.v71i3.2368

Portnoff, L., Gustafson, E., Rollinson, J., & Bicknell, K. (2021). Methods for language learning assessment at scale: Duolingo case study. In *Proceedings of the 14th international conference on educational data mining* (pp. 865–871).

Powell, M. (1970). A new algorithm for unconstrained optimization. In J. Rosen, O. Mangasarian, & K. Ritter (Eds.), *Nonlinear programming* (Vol. 107, pp. 31–66). New York: Academic Press.

Prim, R. (1957). Shortest connection networks and some generalizations. *Bell System Technical Journal*, *36*(6), 1389–1401. doi: 10.1002/j.1538-7305.1957.tb01515.x

Puigcercós, R. M. (2018). Young people' s learning trajectories in the digital age. *Digital Education Review*, *33*, 39–54. (https://revistes.ub.edu/index.php/der/article/view/21560, Last access on 2023-06-09)

Pullan, W. (2009). Optimisation of unweighted/weighted maximum independent sets and minimum vertex covers. *Discrete Optimization*, *6*, 214–219. doi: 10.1016/j.disopt.2008.12.001

Qian, C., Yu, Y., Tang, K., Yao, X., & Zhou, Z. H. (2019). Maximizing submodular or monotone approximately submodular functions by multi-objective evolutionary algorithms. *Artificial Intelligence*, *275*, 279–294. doi: 10.1016/j.artint.2019.06.005

Ramirez, J.-D., Hernández-León, E., Figueroa-Sandoval, B., & Aillon-Newman, M. (2018). Learning and experience: Aesthetics of multimodal texts in higher education. *Digital Education Review*, *33*, 170–184. (https://revistes.ub.edu/index.php/der/article/view/21593, Last access on 2023-06-09)

Regina, D., & Anitha, D. (2022). Computer-based vocabulary learning in the english language: A systematic review. *Theory and Practice in Language Studies, 12*(11), 2365–2373. doi: 10.17507/tpls.1211.17

Rice, C., & Tokowicz, N. (2020). A review of laboratory studies of adult second language vocabulary training. *Studies in Second Language Acquisition, 42*(2), 439–470. doi: 10.1017/S0272263119000500

Ritter, S., & Fancsali, S. (2016). Mathia x: The next generation cognitive tutor. In *Proceedings of the 9th international conference on educational data mining* (pp. 624–625).

Rohrer, D., & Taylor, K. (2006). The effects of overlearning and distributed practise on the retention of mathematics knowledge. *Applied Cognitive Psychology, 20*, 1209–1224. doi: 10.1002/acp.1266

Rosenblatt, L. (1988). Reading and writing: The transactional theory. In J. Mason (Ed.), *Reading and writing connections.* Newton, MA: Allyn and Bacon.

Roy, K., Kar, S., & Das, R. N. (2015). Chapter 6 - selected statistical methods in qsar. In K. Roy, S. Kar, & R. N. Das (Eds.), *Understanding the basics of qsar for applications in pharmaceutical sciences and risk assessment* (p. 191-229). Boston, M.A., USA: Academic Press. doi: 10.1016/B978-0-12-801505-6.00006-5

Saini, M. K., & Goel, N. (2019). How smart are smart classrooms? a review of smart classroom technologies. *ACM Computing Surveys, 52*(6), 1–28. doi: 10.1145/3365757

Sandy-Hanson, A. E. (2006). *A meta-analysis of the impact of computer technology versus traditional instruction on students in kindergarten through twelfth grade in the united states: A comparison of academic achievement, higher-order thinking skills, motivation, physical outcomes and social skills* (Doctoral dissertation, USA). doi: 10.5555/1292855

Sarsa, S., Leinonen, J., & Hellas, A. (2022). Empirical evaluation of deep learning models for knowledge tracing: Of hyperparameters and metrics on performance and replicability. *Journal of Educational Data Mining, 14*(2). doi: 10.5281/zenodo.7086179

Schmitt, N. (2019). Understanding vocabulary acquisition, instruction, and assessment: A research agenda. *Language Teaching, 52*(2), 261–274. doi: 10.1017/S0261444819000053

Schneider, V., Healy, A., & Bourne, L. (2002). What is learned under difficult conditions is hard to forget: contextual interference effects in foreign vocabulary acquisition, retention, and transfer. *Journal of Memory and Language*, *46*(2), 419–440. doi: 10.1006/jmla.2001.2813

Shaffer, D., Collier, W., & Ruis, A. (2016). A tutorial on epistemic network analysis: Analyzing the structure of connections in cognitive, social, and interaction data. *Journal of Learning Analytics*, *3*(3), 9–45. doi: 10.18608/jla.2016.33.3

Shaffer, D., Hatfield, D., Svarovsky, G., Nash, P., Nulty, A., Bagley, E., ... Mislevy, R. (2009). Epistemic network analysis: A prototype for 21st – century assessment of learning. *International Journal of Learning and Media*, *101*(6), 102512. doi: 10.1162/ijlm.2009.0013

Shaffer, D., & Ruis, A. (2017). Epistemic network analysis: A worked example of theory-based learning analytics. In C. Lang, G. Siemens, A. Wise, & D. Gašević (Eds.), *Handbook of learning analytics* (pp. 175–187). Piscataway, NJ, USA: Society Of Learning Analytics Research. doi: 10.18608/hla17.015

Shalin, H.-J. (2018). Applying decision trees to online learning data. *C2C Digital Magazine*, *1*(8). (https://scholarspace.jccc.edu/c2c_online/vol1/iss8/15)

Shi, Y., Xie, W., Xu, G., Shi, R., Chen, E., Mao, Y., & Liu, F. (2003). The smart classroom: Merging technologies for seamless tele-education. *IEEE Pervasive Computing*, *2*, 47–55. doi: 10.1109/MPRV.2003.1203753

Sinnayah, P., Rathner, J., Loton, D., Klein, R., & Hartley, P. (2019). A combination of active learning strategies improves student academic outcomes in first-year paramedic bioscience. *Advances in Physiology Education*, *43*, 233–240. doi: 10.1152/advan.00199.2018

Smiley, P. A., Buttitta, K. V., Chung, S. Y., Dubon, V. X., & Chang, L. K. (2016). Mediation models of implicit theories and achievement goals predict planning and withdrawal after failure. *Motivation and Emotion*, *40*(6), 878–894. doi: 10.1007/s11031-016-9575-5

Sottilare, R., Long, R., & Goldberg, B. (2017). Enhancing the experience application program interface (xapi) to improve domain competency modeling for adaptive instruction. In C. Urrea, J. Reich, & C. Thille (Eds.), *Proceedings of the fourth acm conference on learning @ scale* (pp. 265–268). New York, NY, United States: Association for Computing Machinery. doi: 10.1145/3051457.3054001

Suthaharan, S. (2016). Support vector machine. In *Machine learning models and algorithms for big data classification: Thinking with examples for effective learning* (pp. 207–235). Boston, MA: Springer US. doi: 10.1007/978-1-4899-7641-3_9

Sweller, J. (2011). Cognitive load theory. *Psychology of Learning and Motivation, 55*, 37–76. doi: 10.1016/B978-0-12-387691-1.00002-8.

Tagashira, K., Kida, S., & Hoshino, Y. (2010). Hot or gelid? the influence of l1 translation familiarity on the interference effects in foreign language vocabulary learning. *System, 38*, 412–421. doi: 10.1016/j.system.2010.03.015

Takumi, S., & Masaki, U. (2022). Analytic automated essay scoring based on deep neural networks integrating multidimensional item response theory. In *Proceedings of the 29th international conference on computational linguistics* (p. 2917–-2926). Gyeongju, Republic of Korea: International Committee on Computational Linguistics.

Thomas, J. W. (1980). Agency and achievement: Self-management and self-regard. *Review of Educational Research, 50*(2), 213–240. doi: 10.3102/00346543050002213

Tian, Y., Huang, W. N., Zhang, L., & Wang, T. (2021). Disclosing personal names in screen names predicts better final achievement levels in massive open online courses. *IEEE Access, 9*, 50926–50938. doi: 10.1109/ACCESS.2021.3069451

Tinkham, T. (1993). The effect of semantic clustering on the learning of second language vocabulary. *System, 21*(3), 371–380. doi: 10.1016/0346-251X(93)90027-E

Tinkham, T. (1997). The effects of semantic and thematic clustering on the learning of a second language vocabulary. *Second Language Research, 13*(2), 138–163. doi: 10.1191/026765897672376469

Tokac, U., Novak, E., & Thompson, C. G. (2019). Effects of game-based learning on students' mathematics achievement: A meta-analysis. *Journal of Computer Assisted Learning, 35*(3), 407–420. doi: 10.1111/jcal.12347

Turner, L., Ehrgott, M., & Hamacher, H. W. (2015). On the generality of the greedy algorithm for solving matroid base problems. *Discrete Applied Mathematics, 195*, 114–128. doi: 10.1016/j.dam.2014.08.034

Walsh, M., Krusmark, M., Jastrembski, T., Hansen, D., Honn, K., & Gunzelmann, G. (2023). Enhancing learning and retention through the distribution of practice repetitions across multiple sessions. *Memory & Cognition, 51*, 455–472. doi: 10.3758/s13421-022-01361-8

Waltz, R., Morales, J., Nocedal, J., & Orban, D. (2006). An interior algorithm for nonlinear optimization that combines line search and trust region steps. *Mathematical Programming*, *107*(3), 391–408. doi: 10.1007/s10107-004-0560-5

Waring, R. (1997). The negative effects of learning words in semantic sets: A replication. *System*, *25*(2), 261–274. doi: 10.1016/S0346-251X(97)00013-4

Wegener, S., Beyersmann, E., Wang, H., & Castles, A. (2022). Oral vocabulary knowledge and learning to read new words: A theoretical review. *Australian Journal of Learning Difficulties*, *27*(2), 253–278. doi: 10.1080/19404158.2022.2097717

Welsh, D. (2010). *Matroid theory*. Mineola, N.Y.: Dover Publications.

West, D. (2001). *Introduction to graph theory (2nd ed.)*. N. J.: Prentice Hall.

Whitney, H. (1935). On the abstract properties of linear dependence. *American Journal of Mathematics*, *57*(3), 509–533. doi: 10.2307/2371182

Wilcox, A., & Medina, A. (2013). Effects of semantic and phonological clustering on l2 vocabulary acquisition among novice learners. *System*, *41*, 1056–1069. doi: 10.1016/j.system.2013.10.012

Willis, L.-D., & Exley, B. (2018). Using an online social media space to engage parents in student learning in the early-years: Enablers and impediments. *Digital Education Review*, *33*, 87–104. (https://revistes.ub.edu/index.php/der/article/view/21702, Last access on 2023-06-09)

Xie, L., Chen, Y., & Li, M. (2020). Convolution forgetting curve model for repeated learning. In *International conference on artificial intelligence and education* (pp. 447–454). Tianjin, China.

Xu, Z., Zhao, Y., Zhang, B., Liew, J., & Kogut, A. (2022). A meta-analysis of the efficacy of self-regulated learning interventions on academic achievement in online and blended environments in k-12 and higher education. *Behaviour & Information Technology*, 1–21. doi: 10.1080/0144929X.2022.2151935

Yang, K., & Chen, H. (2021). What increases learning retention: employing the prediction-observation-explanation learning strategy in digital game-based learning. *Interactive Learning Environments*, *31*(6), 3898–3913. doi: 10.1080/10494820.2021.1944219

Yarborough, C. B., & Fedesco, H. N. (2020). *Motivating students, Vanderbilt University Center for Teaching.* (https://cft.vanderbilt.edu/guides-sub-pages/motivating-students/, Last access on 2023-01-13)

Yousefi, M., & Biria, R. (2018). The effectiveness of l2 vocabulary instruction: a meta-analysis. *Asian-Pacific Journal of Second and Foreign Language Education*, *3*(21). doi: 10.1186/s40862-018-0062-2

Yu, A., & Trainin, G. (2022). A meta-analysis examining technology-assisted l2 vocabulary learning. *ReCALL*, *34*(2), 235–252. doi: 10.1017/S0958344021000239

Yves, R., & Frédéric, V. (Eds.). (2010). *Introduction to scheduling.* Boca Raton, Florida, USA: Taylor & Fancis Group. doi: 10.1201/9781420072747

Zapata-Rivera, L., & Larrondo Petrie, M. (2018). xapi-based model for tracking on-line laboratory applications. In J. Rhee (Ed.), *Ieee frontiers in education conference (fie)* (pp. 265–268). New York, NY, United States: IEEE. doi: 10.1109/FIE.2018.8658869

Zhang, Y., Li, X., Zhu, L., Dong, X., & Hao, Q. (2019). What is a smart classroom? a literature review. In S. Yu, H. Niemi, & J. Mason (Eds.), *Shaping future schools with digital technology: An international handbook* (pp. 25–40). Singapore: Springer Singapore. doi: 10.1007/978-981-13-9439-3_2

Zhang, Y., Tiňo, P., Leonardis, A., & Tang, K. (2021). A survey on neural network interpretability. *IEEE Transactions on Emerging Topics in Computational Intelligence*, *5*(5), 726–742. doi: 10.1109/TETCI.2021.3100641

Zhu, G., Xing, W., Costa, S., Scardamalia, M., & Pei, B. (2019). Exploring emotional and cognitive dynamics of knowledge building in grades 1 and 2. *User Modeling and User-Adapted Interaction*, *29*(4), 789–820. doi: 10.1007/s11257-019-09241-8

Zou, D., Huang, Y., & Xie, H. (2021). Digital game-based vocabulary learning: Where are we and where are we going? *Computer Assisted Language Learning*, *34*(5-6), 751–777. doi: 10.1080/09588221.2019.1640745

Index

Kolibri 360 GmbH
Ferdinand-Jühlke-Straße 7
99095 Erfurt, Deutschland
produktsicherheit@kolibri360.de